SWITCH

SWITCH

EXPLORING 'TRIPS' BETWEEN LOOPS AND CIRCUITS IN THE DYNAMIC HUMAN BRAIN

STEWART SANSON

FIRST EDITION

ISBNs:
Paperback: 978-1-80227-636-7
eBook: 978-1-80227-637-4

for Judith, Alexander and Clare

Contents

Acknowledgements

I thank my son, Alexander, for close reading of innumerable drafts followed by numerous and severe suggestions for improvements.

He also drew the diagrams, and prepared the cover, courtesy of Pixabay.

Preface

You are in your office and intend to take a page to the photocopier down the corridor and on the way back give the copy to a colleague in another office. Suddenly you 'come to' in your own office and realize that you still have the copy in your hand, having forgotten to give it to your colleague.

It's obvious what's happened. The repeated acts of going to the photocopier and then straight back to your office have established what I call an automatic motor loop in your body. This loop overpowers *another* less permanent loop or intention which formed your antecedent; conscious intention to return to your office via a colleague's office.

Another observation: why is it when approaching a stationary escalator you stumble slightly whereas you would have thought it would have been easier to get onto a stationary device than a moving one? It's clear that the repeated acts of adjusting your gait as you engage with a moving escalator – a slightly larger lurch forwards, which would otherwise be risky, to land correctly on the mounting platform - have established a visuomotor, subconscious loop in your body but your visual system has returned a misleading signal to your awareness at that moment. You feel a judder and your experience is noticeably modified. I would confidently predict that this perturbation would not be found in those whom never or very rarely use escalators.

These switches then impel noticed transitions between states of conscious self-awareness without me being obliged to say what consciousness is exactly.

At this point, some readers may interpose the so-called 'hard problem of consciousness'; that is how qualities like colour and taste are instantiated in the brain. I cannot explain it but this book does expose some types of qualities, or 'qualia', that *may* be explicable.

* * *

Some time ago I started a folder called 'Switch', half conscious of an interesting phenomenon. Imagine my pleasure when I saw recent scientific papers citing this very notion with the same word. (More in the notes.)

Karl Deisseroth is a neuroscientist who has drawn on his own experience as a father of small children. 'It changes you,' he says:

> … the structure is there waiting for the *switch* to trip. (My italics)

When new parents talk of being 'rewired' or 'reprogrammed' after birth, they are being more accurate than they realize.

Let me give a slightly different but relevant line of thinking in order to convey the essence of what I am going to argue.

Most neuroscience texts on vision show unidirectional nerve paths from the retina, a half-crossover at the optic chiasma, the optic tracts to the lateral geniculate nucleus (hereafter LGN) and the optic radiation to the visual cortex at the back of the brain. No doubt this helps the student as she embarks on the fearfully complex study of the brain. (You may wish to look at the notes for Chapter 1 for a diagram.)

But these diagrams are often incomplete.

Half the input into the LGN is fed in a *loop* forward from the visual cortex; *recurring* loops are established. I will try to explain why this is important.

And a final example: you reach deliberately for a cup of coffee, or perhaps less consciously for another nibble at that bar of chocolate. You think it is all one action.

It is not.

There are numerous systems and subsystems governing these seemingly simple operations, and usually it is impossible to ascertain yourself which pattern you are in.

As we explore these systems I want to build a more nuanced picture of the brain than is conveyed in some neuroimaging studies you may have seen with parts of the brain 'lit up' when it is engaged in a given task.

There is another complication we should be aware of because we evolved from a fish-like tetrapod that hauled itself out of the sea, adapting its breathing mechanism to derive more energy from oxygen, and learnt to walk on land.

That creature evolved into a mammal with striking differences in the parturition and care of its young. Then we find increasing brain size in proto-humanoid apes, then us.

But … it is as if our bigger brains sit 'on top of' an older 'reptilian brain' that governs many essential functions like homeostasis (keeping the physiology of the body stable), metabolism and mating, fighting or fleeing).

Again, we may not be sure which system, or loop, we are in the grip of. Worse, it is possible that this reptilian brain may play a part in our fixed habits even as the more flexible, newer brain, the neocortex, fights for overall control.

However, we do not have to posit two brains, an older one and a new one. Different and overlapping loops between these parts are established but these loops can be hard to break out of.

* * *

I do not claim complete originality for my 'loop' idea. Students of neuroscience will recognize Hebb's explanation of 'cell assemblies' which can be activated by a stimulus and may continue reverberating when the stimulus ceases. However, I make a bolder claim by correlating shifts of awareness with the interactions between loops, in and between both cognitive and emotional circuits, and I add colour and detail as I explore personal experience, the arts and a little science.

In short, the switches and trips in brain, and body, provoke unexpected feelings of an elusive, special kind of feeling, an enhanced sense of 'self-awareness'. These fairly abrupt transitions come into being and arise into awareness, but there is nothing mysterious or inherently inexplicable about them. They are a natural product of evolution; I leave it to you to judge whether they convey a selective advantage according to classical Darwinian theory or, as also can happen in evolution, result in unexpected and less adaptive outcomes.

Or both.

Chapter 1

De profundis

Awareness or *self-awareness* is an attribute of the alert, awake brain but, awkwardly for us, can be fudged with the term 'consciousness' which I shall try to avoid. Any notion of a dichotomous distinction between being aware and not being aware is not helpful either; there are shades of alertness and awareness. That is in part the point of this book.

For some, being aware extends to being aware that you are aware but in my view this is a hazardous route to venture on because it could lead to an infinite regression. Nevertheless, I hope you are comfortable with the idea that you may be aware of a cup of coffee in front of you or that you are aware that you are not asleep. But please look at the notes where this point is amplified because, for example, we are not sure sometimes whether we are in a dream or not.

There is a lot of controversy around the notion of 'consciousness' and I do not mean to dismiss it casually but can say briefly that a few philosophers allege that it is inexplicable; some say that it is a fictive construct; worse, they allege, we are simply muddled.

This book takes a less extreme view, mostly because it seems obvious to me that strange experiences can subjectively be felt.

And so I say that awareness is a natural phenomenon arising from evolution, and there is no reason why it cannot be explained however long it takes and difficult though it may be.

It is an outcome of the quality of a transition between states of physical matter in the brain.

This first chapter introduces my idea about how the feeling of self-awareness arises from what I will call 'switches' between and within 'loops', and then explores more themes embroidered in later chapters.

I outline my idea straightaway so that you have a sense of where I am going, and then I ask you please to nestle each new example within this overall scheme:

> **Fairly sudden switches between parallel loops, or even within loops, bring about within us experiences of altered states of awareness.**

To define some of these terms:

A *loop* is a neuronal or other physiological path, or a *circuit* where elements of the data streams they carry are looped back into themselves but into and from which there are entrances and exits.

And I want to use another term: a 'parallelism', where there are two or more loops or circuits operating at more or less the same time.

I cannot venture to estimate how many loops and parallelisms are operative at any given moment nor how they would all be confidently detected. It will also depend on the context: immersed in an intense game of tennis it may be several dozen; in daydreaming or proper dreaming it could be hundreds. Below you will see a reference to 100,000 fibres in a nervous system which monitors vital functions of the body. The number gives us a clue as to why there are unconscious and conscious processes: we could not possibly hold 100,000 loops in our heads consciously! Instead, kindly nature allocates numerous functions to an automatic sub-conscious system so allowing us in a state of awareness to concentrate on what we think are more important matters – but about which we could be completely wrong.

Loops may be tenuous, created in seconds and subject to disappearing in as much time, but they may become more stable and enduring through both repetitive excitation or sudden trauma. We all forget where we put our keys last; no noticeable loop has been established. The brain did not need to make this effort. But we can all remember a sight we have seen only once, momentarily, and never forgotten.

Switches or *trips* are what the terms imply: fairly sudden changes between states of the brain *which are noticed* as they intrude into subjective experience, and may be held there transiently, or re-looped into longer-term memory and so be susceptible to later recall. The switches are often *reportable* although it would be better to use the passive voice, 'which are noticed' and not 'which we notice', because that would mean we have to define what 'we' is.

The loop will be instantiated principally in neuronal circuits and the interconnections between them at junctions called synapses but it would be prudent to mention that we should not look only at neurons. In Chapter 3 you will come across a hormonal loop, and in the notes I list other features of the human brain that may be implicated.

Loops may be mostly confined to one hemisphere of the brain or cross between hemispheres.

Loops and parallelisms may be *stratified*, not literally in layers but in effect; some stronger, some weaker, some more likely to emerge into awareness, others less so, but even the latter are intrinsically capable of being fortified by their very inter-connectedness with other loops so that working together they are able to induce a switch and make their content available to awareness.

Now to explain the title of this chapter.

I had been struck by a marvellous metaphor penned by the French novelist Marcel Proust:

> Then for a second time I create an empty space before it [an unknown state], I confront it again with the still recent taste of that first mouthful and I feel something quiver in me, shift, try to rise, something that seems to have been unanchored at a great depth; I do not know what it is, but it comes up slowly; I feel the resistance and I hear the murmur of the distances traversed... Will it reach the surface of my limpid consciousness – this memory, this old moment which the attraction of an identical moment has come so far to summon, to move, to raise up from my very depths?

* * *

Retaining the notion that a parallelism can be manifested as information encoded in circuits spontaneously across the brain, let us look at an example.

Working with mice, Stringer and others showed in a 2019 paper that neuronal populations in the sensory cortex, which is in the upper back part of the brain, produce variable responses to sensory stimuli and show intricate spontaneous activity. But at the same time the information encoded *nearly everywhere* in the brain is combined into variable states of behaviour. In Singer's paper we have a vivid parallelism: many different but related and interconnected actions taking place everywhere in the brain.

To quote:

> The most appropriate behavior for an animal to perform at any moment depends on the combination of available sensory data, ongoing motor actions, and purely internal variables such as motivational drives. Integration of sensory inputs with motor actions must therefore occur somewhere in the nervous system.

Let's unpack this.

These memories, which may activate responses, are stored as patterns of connectivity in many areas of the brain, contrary to what scientists had thought for a long time. This was partly because it seemed to make more sense that memories were stored in one place but partly because imaging equipment was not powerful enough to show more attenuated signals from other parts of the brain. Even at the time of writing, 2022, imaging studies have to be combined with pinpoint accurate methods of measurement such as inserting electrodes very carefully into individual neurons but at the cost of not being able to see the wood for the trees*.

> *You can stimulate a single neuron to elicit a memory or behaviour but it does not mean that the memory or behaviour is encoded only in that one cell.

Emily Underwood, in a recent feature in the journal *Science*, uses the term *interoceptive* that encompasses two-way communications within a complex, body-wide system of nerves and hormones. The *vagus* nerve (the tenth cranial nerve for the anatomist) is:

> … a massive meandering network of more than 100,000 fibers that travel from nearly every internal organ to the base of the brain *and back again*. (My italics)

She hazards that these effects may even construct our sense of self.

And she makes another point:

> Many signals are unconscious, and when they do rise to our awareness, they're often blurry and ambiguous, making it hard for people to report what they are experiencing.

This is exactly what I think, with the qualification that the experience could also be salient, unmissable, reportable, and sometimes frightening.

Most of the time the vagus nerve gets on with its job, well out of the reach of awareness. But, as one example Underwood gives, suppose you over-indulge yourself at dinner and there is tug in your chest. That comes into awareness and:

> your brain has to decide if it's because you ate too much – or it is the early sign of a heart attack.

I would call this 'hierarchical loop sorting'; you are using your knowledge to measure the feeling against what your body is telling you. And one more recent paper here by Toker and others from the 'Significance':

> What happens in the brain when we lose consciousness? One possibility is that the loss of consciousness corresponds to a transition of the brain's electric activity away from *edge-of-chaos criticality*, or the knife's edge between

> stability and chaos ...We show that the electric activity of the brain is indeed poised near the boundary between stability and chaos during conscious states and transitions away from this boundary during unconsciousness and that this transition *disrupts* cortical information processing. (My italics)

The loop idea touches on an array of our cognitive and other functions so now I want briefly to broach three themes where, in my opinion, distinctive loops are demonstrable: memory, dreaming and attention.

I defer *emotions* to a separate chapter.

Memory

If switches are sometimes so transient, how can we store a persistent memory?

There are numerous ideas but salient are those positing preserved, enduring synaptic patterns of connectivity in nerve cells. For example, if you can play tennis well or ride a bicycle, automatically you have achieved consolidated patterns of motor-sensory loops that do not melt away overnight. Even so they will start falling into desuetude, or less accurate responses, if you do not keep up that tennis practice.

I allocate to the notes a consideration of oscillations at different wavelengths in the brain.

Remembering that the Eiffel Tower is in Paris may not seem quite the same as a motor memory. However, many scientists would say that, in essence, it is the same kind of thing.

Walter Gonzalez and others write:

> The likelihood that a neuron maintains its responsiveness across days is proportional to the number of neurons with which its activity is synchronous. Information stored in the individual neurons is relatively labile, but it can be reliably stored in networks of synchronously active neurons.
>
> A scientific explanation has also to involve a structure called the hippocampus, a key part of the memory system. It holds short-term memory and, depending on the exigencies of context, shifts short-term memories to long-term. It informs you whether an experience is new or has been retrieved from memory.

Let me call it a 'loop sorter'.

Dreaming

Here I believe that my loop idea can explain a lot.

But first, sleeping is not fully understood. It rests the body and brain (to a limited extent) but it is not evident why we have to make a transition into such a different state. Dreaming takes place when we sleep.

Actually, it is not quite so clear-cut. Hypnopompic states are transitions from sleep to waking, and hypnagogic are states immediately before falling asleep.

We also loosely use the phrase, 'daydreaming', which might apply to a child not concentrating on his schoolwork or much more extended, and often pleasant reveries which we know, *in parallel* with a governing master switch, are not grounded in reality but afford us escape from our daily worries.

These oddities do not challenge my argument, rather they enhance it.

Back to dreams: they puzzle us and have evoked much speculation. Do they tell us anything useful about loops and parallelisms?

I think so.

Dreaming could be the result of chaotic loops that randomly intrude and are composed of age-long deeply seated worries mixed up, in no discernible pattern, with the most trivial of happenings of the preceding day. Dreaming takes place when we are sleeping but this means that the loops are disconnected from loops arising from the awake state which provide more solid data, within loops of course, on the position of the body, its interaction with the environment, gravitational clues monitored by the vestibular apparatus of the middle ear, and the results of purposeful, conscious actions.

That can explain the bizarreness of dreams. Totally unconnected wisps of your experiences in the preceding day try to establish an enduring pattern – and fail.

And yet there must be something more solid. Marcel Proust (who suffered from insomnia) asked:

[after leaden sleep] ... how then, searching for our thoughts, our identities as we search for lost objects, do we eventually recover our own self rather than any other?

Loop theory can answer.

On waking, the loops which constitute our sense of self – our identity, our name, our relationships, our biographical past, the name on the demand from the income tax authorities – re-assert themselves forcefully and give us back our own self.

Attention

Within awareness, there is a spectrum of sensitivity to stimuli and intensity of responsiveness. And attention may be switched from one loop to another either by what seems like a voluntary process or by external exigence.

In the day, you will be calling on hundreds – or thousands – of established loops which will help you get through your routines: getting dressed, going to work etc. It is not difficult to put on your shirt and to tie your shoelaces whilst thinking ahead about what to say in that tricky meeting.

Less repeated or novel actions will set up temporary loops instantiated in cerebral tissue. In other words, neurons may be activated in synergy for different lengths of time.

Most of the time we are not paying attention to our being conscious or alert; we simply get on with life, having a meal with friends, washing up, working …, whilst attending to the task in hand. It is not noticed by us that we have this strange brain which sometimes wonders about itself.

In the next chapter, and in much more depth, I tackle loops in the visual system, with its vital, associated systems.

Chapter 2

To see or not to see

The faculty of vision exhibits numerous switches, loops and parallelisms.

For instance, the *ventriloquist effect* is a loop that we can barely escape from.

Even when a less-than-human rag doll is moving its lips and the artificial sounds emitted by the ventriloquist are *in register* with these movements, we have the overwhelming impression that the doll is speaking. I have heard little pigs and all manner of creatures talking. You see many in cartoons and animations.

This is because we are so used to watching the lips of people as they speak, even though it is not completely necessary, that our behavioural response has become ingrained, automatic and looped in.

The effect can be explained by reference to structures called the *colliculi* (little hills). These are two pairs of bumps on the upper part of the top of the midbrain.

In mammals the superior colliculus has three dorsal (upper) layers which primarily deal with visual sensory information. Four deep layers ventral (lower) direct orienting movements. P.J. May says:

> … the motor field organization is in register with the sensory field organization of the overlying superficial layer visual map.

But the superior colliculus also receives auditory information from the *inferior colliculus*, which maps sources of sounds, and associates the two together, thus explaining the ventriloquist effect as a type of cognitive double-loop.

Nobel prize-winner Eric Kandel says that there is:

> a movement map in the colliculus, and this map is in register with the visual map.

The colliculi, taken together, have a visual map plotting external points of reference, a map of its own body's surface, and a map for the direction of sounds in space. These maps are forwarded to motor maps which direct the head and eyes to targets.

Structures in the posterior parietal cortex overlap – in parallelisms – with the superior colliculus as functions for distal vision, e.g. detecting the movement of a predator further away, or proximal vision as in handling objects and tool-making to which I shall come later. Even though, as Livingstone and Hubel say:

> Furthermore, the colliculus seems to be relatively more important in lower mammals than it is in primates.

The function of these structures shows that vision is not simply the handling of visual information; it is part of a system that coordinates vision and other senses with quick automatic movements in response to stimuli – visual, auditory and somatosensory – *and* intended actions directed at precise locations outside the body.

It is worth dwelling on this point.

In the course of evolution some cells destined to evolve into the retina became particularly sensitive to light so that the organism acquired more useful information about its environment. But this mutation would only have been useful – and therefore preserved – if the information so detected was transformed into a beneficial response; in this example, the integration of a proto-visual type of data detection into an appropriate movement, let us say in the case of a worm moving towards or away from light.

Coming up to the present in the human, there are at least three pathways in vision, sometimes interconnected: one is concerned primarily to detect movement but not shape and in poor lighting conditions, one is for the colour and the precise detection of outlines in good conditions, and one is for colour only.

The processes go at different speeds.

From the point of view of evolution, this development is explicable. The detection of predators needs to be done fast, without necessarily being able to see exactly what is moving; if a shadow crosses us fast in our receptive field we flinch. The information has been sent primarily to the midbrain where it can be quickly integrated with movement. On the other hand, later in evolution, judging the ripeness of a fruit by assessment of colour and texture is done more slowly, and for several reasons. One, the analysis at the retinal stage is more complex. Two, the outcome of the processing can be sent to the neocortex for further treatment because the neocortex has more powers of association, so that the results of the analysis may then be integrated with other cognitive faculties and the outcome of this second type of operation may be put into a more permanent form of memory. A chimpanzee will perceive where semi-ripe bananas are found and can return at a time, judged by experience, to the same place to pick them as they ripen. In contrast, our flinching from the movement of a shadow is instinctive. It involves a different part of the brain and there is no need consciously to remember anything.

The cerebellum, the poorly-named 'little brain', is at the back of the head and lower down contains more neurons than the cerebral cortex, and deals with the fine-grained calibration of movements – that is the intended outcome of a muscular movement and readjustment if the outcome does not equal the intention. That is why I call it the 'loop-master' *par excellence.*

The cerebellum also deals with balance, working in intricate, looped partnership with the *vestibular apparatus* for balance connected to the inner ear. It is also implicated in memory.

As long ago as 1988 Margaret Livingstone and David Hubel wrote a paper about the segregation of form, colour, movement and depth. They use a phrase which may be better than my own:

> ... cells in cortical visual areas ... and higher visual areas are segregated into three *interdigitating* subdivisions that differ in their selectivity for color, stereopsis, movement and orientation. (My italics)

They had discovered distinct structures in one layer in the visual cortex which they termed 'blobs' and which are especially sensitive to different wavelengths of light and to brightness, as a function of intensity. This gives us 'colour'. They are not found in lower mammals.

From all these loops and parallelisms the brain must be able finally to 'construct' a coherent, meaningful experience.

The striate* cortex at the back of the brain (V1) is split by a longitudinal fissure in the vertical plane, clearly dividing these parts of the two cerebral hemispheres. Only cerebro-spinal fluid fills the gap. Less noticeable but nevertheless making a clear scission, the occipital lobe is divided in the horizontal plane by the two *calcarine* fissures on either side of the longitudinal fissure into upper and lower banks. This means that the upper part of the right external visual field, for example, is finally projected to the lower part of the left occipital cortex, and the lower quadrant of the left visual field is projected to the upper quadrant of the right cortex, and so on.

> *Striate: so-named because of its visible appearance in layers.

Then how can 'lines' seems so straight, regular and joined up? And if the image is inverted and reversed on the retina, how is it that houses are the right way up and complete?

Our perception of the outline or shape of, say, a book on the table before us, relies on more than one form of detection, thus integrating other loops. It depends on continuous fine-grained ocular movements, touch, and the geometric manipulations of our body to get our fingers to it. A perceived line is not so much an illusion but a 'construction', a statistical 'filling in' of some kind to fulfil laws of probability so that what we know to *be* whole from total experience appears to be *made* whole in experience. The internally assembled construction fits the intrinsic, external characteristics of the object but that does not mean they are the same thing; they only amount to the same thing because we use one to derive the other and both continuously interact. Vitally, the internal construction is 'put' in register with or is aligned with the spatial location of the external objects. And when I say that internal construction relies on more than senses *per se*, I mean that the brain is doing something much more than integrating data from the senses.

The house is the right way up because we know the ground is 'down there' from pressure on our feet or backside as well as from our *idiopathic* sense of which way up our own body is.

I believe scientists have at least explained how nearly spherical cells with processes, which cannot possibly be lined up with the same precision as the straight edge of a material object outside the head, can perceive the orientations of lines, and the movement of lines, edges and curves. What is made of them is another story but temporarily I have to explain my assertion that 'there cannot be lines in the head'. What I mean is that there cannot be 'lines' embedded in cerebral organic tissue in the same way that they are to be found in smooth, straight continuity on objects external to the head. However, 'lines' as such may well be 'made', or seem to be made, or put together somehow in the head. And yet what is in the head is not sufficient. It must be trained by the viewing of external objects and, when 'seeing', be in a continuous receptive mode by the act of looking. The consequence is that in the act of seeing the relative contributions from the external stimulus and from internal cognition may not be calculable. We *switch* constantly between the two even as we are looking.

We – and probably most other mammals – detect, and then in the appraisal, re-detect, and re-appraise in cycles of varying durations depending on the situation as a whole. In other words, looping is an inherent part of how we perceive things.

For one more example, consider *Rubin's Vase*:

For how long can you sustain the continuous perception of two faces or one vase?

Most people switch, involuntarily, between the two. The pattern of black and white on the page or screen remains exactly the same but the brain makes perceptual sense quickly of one or the other interpretation but then, in its restless search to extract meaning from data, realizes an alternative construct is possible.

So, we ignore, or respond, or remember, or forget, or join what we have appraised to other appraisals; these operations being carried out principally in the neocortex. What is beyond question is that in an animal such as the frog no information is sent to the neocortex for the very good reason that it does not have one.

Similarly, different types of analysis are later performed on moving objects. The *where* is as important as the *what*. An enriched form of cognition, and thus 'perception', evolved as the neocortex evolved. Therefore, the subcortical system integrates visual detection with what the *body* is doing, and also installs there a routine – a loop – of muscular memory. This is essential in forming the sense of a self. At the same time the cortical cycles generate a system for perception which may integrate cognition, memory, reason and judgement; all of which will inflect the choices of movement and actions which the person will attempt.

Crucially, the spatio-temporal persistence we accord to an object, as part of the establishment of its reportable identity, is derived from cognitive operations drawing on the spatio-temporal distribution and motions of parts of the body as well as the cognitive operations derived from appraisals of features other than spatial coordinates. The position of the object and the position of the observer-actor are as important as the colour or other features of the object.

It may be asked if it would not have been better to have aggregated all related functions into one, discrete part of the brain.

If starting from scratch, maybe. But evolution does not work like that; it builds on opportunity. So as enhanced abilities developed – were 'selected for' in Darwin's words – then the brain had either to grow new tissue or take over other tissue in a crowded skull for each skill. It might not in the long run have made such a difference. Because of the manifold operations in vision, the tasks have to be spatially distributed in the brain anyway. In an animal there is no such thing as vision only, as there is for a camera or similar electronic device. In the animal, vision must also be coordinated with other functions, like movement and hearing.

There is a penalty. In paying for the mastery of cognitive operations, underpinned by a conviction of the persistence of the visible world, the brain succeeds in deceiving itself. It presents smoothed out continuous appearances to its bodily owner, whereas the actual constitutive elements – in the data streams – of those appearances are highly fragmented.

I propose then that the position of the construction is seemingly re-aligned with the spatial coordinates of the detected object because the brain has to combine knowledge of the perception, that is the construction of the whole thing, with knowledge of the spatial aspects of detection, done largely in a different part of the brain and in a different way. Otherwise, it would be conflicted by heterogeneous information.

The brain is not clever enough to undo its own self-deception in the matter of the actual placement of the perception *and so it outwits itself.*

To say more about these parallel loops:

Two main paths developed from the back of the brain. The mediation of motion and depth proceeds largely by an upper, dorsal path to the posterior, parietal cortex; elements contributing to contrast, shape and colour proceed largely by a lower, ventral pathway to the inferior, temporal cortex. This is the scheme in broad outline for the functions and the main paths, broadly defined:

Upper (*where*):

- *magnocellular* cells
    - motion, dynamic form, depth
    - some colour
    - faster

Lower (*what*):

- *parvocellular* cells
    - predominantly colour, also form and depth
    - slower

In the notes you will find these parallelisms and loops set out in even more detail.

To put a figure on 'faster' and 'slower': signals through the magnocellular path arrive about 20 milliseconds before the parvocellular signals at the striate cortex.

Now for a critical point.

Processing of colour in what will become perceived colour – rather than chromatic contrasts exploited to detect the edges of things for the purpose of detecting moving objects – became lateralized to the *left* hemisphere. The brain evolved somehow to allocate a particular patch of tissue to this job.

Why?

Monkeys and most non-human apes usually prefer one hand for reaching and picking up an object, but it does not matter which. About half of the monkeys studied use the left hand and about half the right hand. The human ape also prefers one hand, especially for skilled manipulations, but it is most commonly the right one. Therefore, we may venture that a fruit-picking or tool-making habit perhaps became associated with the right hand – we do not know why exactly – and then that tool-making skills evolved, often using one and the same hand.

Now, to revert to the upper pathway, the colour of a fast-moving object is not significant if the moving object poses a danger; better get out of the way fast or turn and confront the menace, or give up. But transitions in colour – variations in the composition of electromagnetic radiations – are what help in assessing the ripeness of fruits.

We may call this a transition from one loop to another.

The muscular effort – and feedback – required to control the convergence of the eyes also feeds knowledge into the brain and is part of the performance of detection and then 'seeing'.

If there is feedback from V1 to the LGN which then continues to project its state or parts of it back to V1 then it is clear that a *re-enforcing* loop is set up which sharpens the quality of the data stream.

Attention is part of another parallelism which is often overlooked (so to speak).

The effort to direct the eyes to an object stimulates cells in the superior colliculus, which I have already described. This has been found by training a monkey to observe various stimuli in a wide receptive field and then requiring it to direct attention to one object in order to be able to choose and then respond to a specific feature or quality of that object. Michael Goldberg and Robert Wurtz also found enhanced responses from specific cells in the posterior, parietal cortex, which is known to be involved in attention because damage to this part affects behaviour. What is most intriguing is that some cells respond in the same way whether the monkey made a saccade of the eyes to the object or reached for it.

Luo and Maunsell also say that 'attention' can be subdivided into neurological components corresponding to distinct behavioural effects distinguishing between 'vigilance' over hours or days, exploring a new environment, say, and scanning for enemies, and focussed attention where you consciously undertake a hard reading or by accident something is brought to your attention like, say, you have noticed that your dinner is burning. They identify diverging streams of processing.

They also make a point about the 'lingering' effects of what the subject previously attended to. I shall amplify this point later when I come to literature.

So, the brain can meld into a cellular disposition a new abstract, unified entity; that is to say an 'object-ness' and a 'reach-ness' combined.

It suggests a new way of thinking by which, when we consider the matter in cellular terms, some cells which otherwise in terms of cellular machinery function just like any other cells, are able to instantiate two completely separable and utterly different essences – visual characteristics and internal coordination of motion – into one.

On this view, the positional coordinates give sufficient causal power to facilitate targeted movement; the wavelength responses furnish causal power of a different order, namely the possibility of forming a discriminatory and judgemental decision, whose destination is somewhere else in the brain where different kinds of operations are instigated. With regard to the positional theme, I am not saying that it too cannot be further processed into more cognitive, complex information; I am saying that such operations are not always needed in order to reach for an object. There's more in the next chapter and in more detail about the stupendous act of reaching for a cup of coffee.

With regard to causal power, my idea completely eliminates the problem of how the 'mind', supposedly immaterial, could work on matter and *vice versa*. There is only matter acting on matter.

In short, and to augment my argument, the brain is able to make a better, cleaner 'construction', or internal human experience, of the objects and scenes which its retinas have detected. (The retinas are part of the brain.)

When we lose our focus, we waver between different loops; the brain is trying to make up its mind although objectively considered a single fixed pattern falls on the retinae.

'Blindsight'

This phenomenon is found in certain patients who have suffered some sort of cerebral accident such as a stroke.

They claim not to be able to 'see' certain objects yet when the experimenters ask them to reach for certain objects – which the patients say they cannot see – and manipulate them, they can often manage.

The explanation is that there must be two processing streams in vision – at least two loops therefore – one dealing with what we acknowledge we see and the other a system that subconsciously marries vision to the action of the hands and fingers.

CHAPTER 3

On picking up a cup of coffee

You are deceived if you think this is trivial.

Instead, I am going to argue that reaching for a stone deliberately and doing something significant to it with an implement in the other hand is the single most important advance that humans made.

So, here is a eulogy to a complex process made of loops and parallelisms which also induce a powerful sense of our self-awareness through the manipulations of our hands and the balancing of our torsos.

For the science, I am indebted first to a review by Melvyn Goodale who wrote in 2014 that little attention had been paid to the way in which vision is used to programme and control our actions, particularly the movements of our hands and limbs, and so provides an example of a parallelism (though he does not use that word).

Vision

There are two streams of data processing: one relating to visual *perception* and the other to visual *action*. The first stream is ventral (lower) projecting to the inferior temporal cortex and will be instrumental in the final creation of the common visual experience we have of objects and scenes that we can identify, attach meaning and significance to, and name. This is sometimes called the *what* stream.

The second is the *where* stream and is more dorsal (upper).

These will be familiar from chapter 2.

Goodale explains why there are two separate systems; the answer lies in different computational requirements. But he could have added that they evolved at different times, the action stream before the stream for perception. He says:

> ... the visual information controlling action is often inaccessible to conscious report.

To be able to grasp an object successfully for example, the visuomotor system has to deal with the actual size of the object, and its orientation and position with respect to the hand used to pick it up.

But we also move and so the coordinates for action need to be updated fast in real time.

Although there are two streams it is evident that they work together. External objects are processed in *parallel* by the two mechanisms, each transforming the information for different purposes. So, for example, we identify a suitable tool visually and we then grasp it correctly.

Now for more on the *what* stream: when we say that it identifies what an object *is*, we have to look at the implications.

It involves memory, which may or may not be salient, so that when we encounter objects again, we:

> ... attach meaning and significance to them and establish their causal relations – operations that are essential for accumulating knowledge about the world.

In my opinion, these two systems do not make for a 'split' personality, rather they build and enforce the sense of the unity of ourselves. Even the more automatic, 'robotic' dorsal stream provides the feedback needed for adjustment and that adjustment can make its way into consciousness. I argue that using tools in a sophisticated way, combined with language, is part of what makes us self-aware beings.

Takai and others have written about the neural basis for hand muscle synergies in the primate spinal cord. By 'synergy' they mean a newer basis for precise control superimposed on an older faculty in the spine – not the brain – where there are spatiotemporal patterns of muscle synergies in primates, the class to which we belong, during a precision grip task. There is an older premotor system which also provides synergistic control of hand movements.

Anatomy

The arm does not stick out of the body like the gantry of a crane. It is articulated in an ensemble of bones and muscles which permit great flexibility of movement, the application of hard accurate force in throwing or beating, and finely controlled grasping and touching as in painting, sewing, other crafts and hammering away at our laptops or swiping on our mobile phones. The arm also contributes to the body's stabilizing system by shifting loads as we walk or prepare to propel objects.

The shoulder girdle muscles hold the assembly together with muscles anchored to the shoulder blades, the scapulae, and rib cage. The scapula is moved to put the head of the arm, the *humerus*, in the best possible position for the action intended. The humerus itself can be moved in a cone-shaped space extending outwards from the body. The two bones from the elbow to the wrist - the *ulna* and *fibia* - are rather more restricted in movement. They can extend or retract the hand but, vitally, they can rotate the wrist through 180°. The wrist itself has some lateral movement and is rather more up and down, depending on which part of the 180° rotation it finds itself in. Using a slightly cramped movement of the humerus and shoulder so that the arm is extended to one side we can rotate the palm by 360° and still oppose all our fingers to the thumb. This posture may not be very useful but if we then bring our hands behind our back, usually palm up, we can pull the ropes of a sledge.

The thumb and each finger can be individually controlled. The thumb can be opposed to the big finger but also to each finger. Only the bonobo chimpanzee is able to oppose the thumb and index finger.

Hominids developed extra strength in the thumb.

Our hands and fingers can adopt a fascinating range of grasps and movements. If you want to know more please refer to the notes for this chapter, and each time you undertake a given task – whether unscrewing the lid of a jam-jar or inserting and tightening a screw – school yourself to stop momentarily and think about precisely what you are doing and whether your fingers or hands have 'run away' with the action.

According to Deloison, our hand is 'primitive'; i.e. not *over*-specialized in its anatomy. In other words, our predecessors did not go along a narrow path that led to a single kind of grasp or other movement, however effective – exhibited, say, by the snatch of a cat with its paw. Instead, it is polyvalent and can be moulded by the brain to perform an astonishing variety of tasks through parallelisms all waiting to be summoned by the task in hand.

And Merlin Donald adds:

> Humans did not simply evolve a larger brain, and expanded memory, a lexicon, or special speech apparatus; we evolved new systems for representing reality.

And this 'reality', which we should consider provisional and relative, is built by the several modes of perception which I have been attempting to delineate and which are integrated to control the hand.

Neuronal control

To reach for or grasp something requires the integration of at least three sets of information and some kind of intention. The body has to know where the object is, where its own arm is and how it is orientated, where its eyes are looking, and what it intends to do. Its intention may be more or less complicated. It may wish to reach for a cup of coffee placed in its customary position on the table beside the laptop. In this case the brain simply has to select a movement from its repertoire of movements and execute it. We all know how easily the hand stretches forth for that remaining bar of chocolate when our intention had been only to consume part of it. On the other hand – to avail oneself of a useful expression – we might have to plan a movement with one or both hands, like sharpening a pencil, by holding the pencil-sharpener in the left hand and then pushing in and rotating the pencil with the right.

When we reach for an object the arm is usually first projected 'ballistically', that is 'thrown' out by a stereotypic movement largely controlled by part of the brain called the basal ganglia and calibrated by the cerebellum, which corrects for inaccurate movements (and, as I said, has more neurons than the cerebral cortex).

Then we guide our hand and wrist more carefully to the target using our eyes. You may be able to sense the transition between these two methods of control by putting an object in front of you, quickly reaching for it and then putting your finger on a spot on the object. In my terms we have a succession of loops of which we are barely conscious.

The specific motor areas are in the same lobe as the prefrontal cortex, the lobe which shows the most activity during 'higher' cognitive processes.

Whole-arm movements may be considered as operating in an intrinsic frame of reference, deriving substantial information from the body's knowledge of where its parts are in addition to the visual location of the target. By contrast, an extrinsic frame of reference is invoked when particular wrist movements – once the hand has got to its target – are related to the manipulations required as prompted by the nature of the target and what has to be done to it.

Frank Wilson has written a wonderful book, *The Hand – How its use shapes the brain, language, and human culture.* It is about much more than this limb. In a part dedicated to *The Arm We Brought Down* (from the trees). He says:

> st have responded to this purely biomechanical change (from the brachiating or tree-dwelling apes) by increasing the complexity of its representation of the arm and hand in space ... there is no division or segregation of these functions in the body itself; the neuromuscular system functions in a fully integrated way, so that the overall movement is unitary and fluent. In reality, shoulder, arm, and hand functions are exquisitely tuned and responsive to one another in both neuromuscular and biomechanical terms. The kinetic and informational processes take place simultaneously from the body outward to the hand, and from the hand inward to the body.

The proprioceptive control and information systems of the body allocate a large share of their resources to hand control. Furthermore, because of the continuous use of our hands and a heavy investment in training them to perform complex tasks, the space they define in their operations constitutes an extension of the domain of egocentric awareness. In other words, our bodily awareness, mostly defined by the 'space-suit' which allows us to live on dry land and which is partly defined by the skin and partly by internal homeostatic, monitoring devices is extended by about 50 cm in an arc in front of us as we manipulate objects.

Accordingly, this argument counters those thinkers who believe that there is no 'self'. There may not be a metaphysical self, an indelible person which endures even beyond death, but the complex use of tools and all the other appurtenances that modern civilization has created must of necessity lead to the construction of a coordinating principle, which we should not seek in one particular area of the brain and body but in a network of interrelationships.

For a long time I searched for a mechanism that might support this notion and recently found one in a paper on tool use and language by Thibault and others in the journal *Science*. Using functional magnetic resonance imaging when the subjects engaged either in syntactical analysis or in using tools, they discovered that small portions of the basal ganglia in the human brain act as common neural substrates for both tool use and syntax in language.

Picking up that cup of coffee

Back to the cup of coffee when you use multiple functions to pick it up.

The *what* and *where* streams, constituted by separate loops in parallel with each other, are coordinated.

Memory is involved because you put that cup there yourself (or you are able to recollect that someone brought it for you).

There will be more continuous cognitive engagement because you must know how hot the coffee is likely to be, and its degree of depletion – because this affects the mass that has to be lifted – and how exactly you are going to grasp it.

Guided by the *where* system, in parallel, the body locates the cup more precisely to grasp it, at the same time preparing a suitable configuration of the wrist and fingers.

***

Our remote ancestors did not drink coffee from cups though they probably drank water from some kind of container.

What is more important is they were learning to make tools. This ability, which evolved over millions of years, laid down functions that now help us lift that cup of coffee with ease. And without requiring millions more years this highly adaptable capacity, after a period of training to be sure, allows us to play tennis or wire a plug.

Australopithecines, 'Southern monkeys' like the famous 'Lucy', used simple tools like a stone from which a flake had been struck off to provide a cutting edge and could have been handled to butcher a carcass or sever the roots of tubers.

Early hominins, as we must now call them, could have made some tools from sticks and bamboo, of which no evidence remains. Firm evidence shows that two and a half million years ago shaped stone tools appeared in Africa during what is called the Oldowan culture, from Olduvai in the Great Rift Valley of eastern Africa. They consist of stone cores with flakes chipped off to produce scrapers and choppers. Another point: it takes two hours of concentrated effort to make a bifacial axe provided you do not make a mistake.

Some stones may have been used for throwing.

From the scatter of stones between early settlements it can be inferred that the hominins carried stones from one locality to another, which also implies forward planning. I am not dwelling long on this point but add that with the making of objects concepts of ownership and exchange were engendered.

***

In brief, our manipulation of objects and our cognitive resonance with them afford numerous instances of loops and parallelisms which may or may not be foregrounded in an overt conscious or self-aware experience.

Chapter 4

Green-ey'd monster

Now to explore switches in another physiological category.

There are so many different *emotions*: happiness, discontent, jealousy (Shakespeare's 'green-ey'd monster'), anger, fear.

We should distinguish them from feelings like breathlessness, hunger, thirst, being too hot or too cold, and the need for relief in the rest-room, which are all more basic physiological responses. However, the five emotions listed just above are also physiological processes but with added complications.

Take hunger in the physiological category, noting that switches in physiological processes are numerous and practically continuous as the body maintains homeostasis, or equilibrium, and, partly for this reason, are delegated to the subconscious.

Remember from Chapter 1 that *switches* or *trips* are what the terms imply: fairly sudden changes between states of the brain which are noticed as they intrude into subjective experience, and may be held there transiently, or re-looped into longer-term memory and so be susceptible to later recall. The switches or loops may be *reportable.*

The body 'knows' when it is running out of energy supplied in the blood and from the liver. You are not likely to detect this switch. However, you may detect the next one when the body automatically switches from acquiring more energy from fat deposits and muscle tissue. Then, I posit, there is a 'third-level switch' when you sense the unmistakable pangs of hunger and are driven to do something about it.

Control of your body temperature, a vital function, offers another example of a switch. The body constantly makes minor adjustments by sweating or control of water balance. The argument here is that the switch can be detected, breaking into awareness, making you search for a pullover or visit the toilet.

But a switch may be provoked not only by a physiological process or response to immediate stimuli but a cognitive appraisal that there is a danger – or perceived risk of some kind. For example, you are opening an envelope from a hospital where you went for a test or scan. Or it is dawning on you that your marriage or partnership is not working. You have hidden this assessment from yourself but a salient incident throws a switch and, with an emotional jolt, you realize you have to contemplate the end of the relationship.

Two loops exemplified

The notes show a couple of loops from a standard undergraduate text book.

You can see that in the feedback loops excessive cortisol suppresses the secretion of the hormone from the hypothalamus that starts the loop. Cortisol, which is a stress hormone in the bloodstream, then decreases.

The scientific study of the emotions has proven to be particularly difficult. We all know what they are and have experienced most of them. But they are hard to define and measure with precision. Where the scientist prefers objective, repeatable measurements, she or he is confronted by changeable processes and complex systems which, to our subjective grasp, range from barely perceptible moods to violent rages and all-consuming passions.

And, it is especially difficult, by *verbal description*, to convey gradations of feelings although their onsets are usually more perceptible. Nor can we be perfectly certain about what emotions are driving us at any given moment.

> However far a man may go in self-knowledge, nothing however can be more incomplete than his image of the totality of *drives* which constitute his being. He can scarcely name even the cruder ones: their number and strength, their ebb and flow, their play and counterplay among one another, and above all the laws of their *nutriment* remain wholly unknown to him. This nutriment is therefore a work of chance … (Nietzsche)

Following that philosopher, let me quote from American neuroscientist Joseph LeDoux:

> Emotions are notoriously difficult to verbalize. They operate in some psychic and neural space that is not readily accessed from consciousness … the coin of the evolutionarily old unconscious realm is nonverbal processing.
>
> … emotions did not evolve as conscious feelings. They evolved as behavioral and physiological specializations, bodily responses controlled by the brain, that allowed ancestral organisms to survive in hostile environments and procreate.

Both are saying the same thing in different ways. The point is that the emotions, along with numerous other interior physiological processes lacking the power to break into awareness – which is when, according to me, switches happen – are working all the time in the body unknown to us. Only when the results of fear, say, and its associations, depending on the circumstances, force themselves into awareness, do we understand them better, or at least are able to make an attempt to describe them. The awareness in this case is not necessarily the bestowal of new cognitive ability but an awareness of bodily changes: sweating, breathing harder and wrenches in the gut.

What are the emotions actually for when hate, envy and unhappiness are so often destructive? A biological perspective gives an answer for we can be sure that the emotions would not have evolved in animals, including man, if they had served no purpose. Here I have hinted at a mechanism for self-preservation but I also suggest that, within an appropriate cultural setting, aggression for example can be channeled into determination in the face of adversity, and fear into the desire to overcome obstacles through the acquisition of knowledge.

In John Milton's epic *Paradise Lost*, after Satan had been hurled headlong from the 'happy Realms of Light' and left confronting 'Ages of hopeless end', he expresses a defiance and determination which a soldier in retreat could admire:

> What though the field be lost?
> All is not lost, th'unconquerable Will,
> And study of revenge, immortal hate,
> And courage never to submit or yield.

The emotions, often thought of as inconvenient saboteurs of our rational existence, drive creativity, love, endurance and social cooperation. They make us what we are as surely as the rational powers in which we invest so much confidence.

LeDoux's ideas go like this: animals a long time ago and those still living wild today have to be able to respond very quickly to danger.

Fear is an innate mechanism that drives an appropriate response. Milliseconds become a matter of life and death; the response has to be made unconsciously and is usually stereotyped, driven by instinct to freeze or run. If we think about our hominin ancestors, a mechanism that allowed analysis of the danger meant that they learned to slow down their response just long enough to find out what the danger was, marry it to past experience, and respond in a way that was better adapted to the situation. We do this most of the time as we learn that we cannot just hit out at someone who has made an offensive remark to us; it is likely to be counterproductive and there may be a better way of responding. 'The brain buys time', says LeDoux. The hominin was reacting in a different tempo to that of a reflex, by concentrating, whilst still being able to think very rapidly. This gave him an edge over most other animals.

Instantiation of fear

An essential part of the brain involved is called the amygdala (although another site called the insular cortex is also involved). It receives information, usually visual or auditory, from the thalamus by a short path. It is a relay whose job is to get the autonomic nervous system activated fast, and it does not therefore analyze the stimulus. But in the course of evolution the comparatively slower links to the sensory cortex and then the association cortex strengthened. This brought about meaning in the context of a drive for survival and therefore greater understanding. Man is not always the victim of stereotypic reflex action although it is still there. Thus, someone going deliberately to a bridge to throw himself over it and kill himself would automatically jump sideways if he detected something falling on him.

LeDoux says:

> The cortex's job is to prevent the inappropriate response rather than to produce the appropriate one ... those animals that can readily switch from automatic pilot to willful control have a tremendous extra advantage. This advantage depends on the wedding of emotional and cognitive functions.

He points out that the amygdala was present in dinosaurs eons ago, in fact in all vertebrate brains because detecting and responding to danger was so primordial. That part of the brain has not changed much – 'we are emotional lizards'.

The emotional responses provoked in the body (sweating and increased heart-rate) are then brought back into awareness through the cortex and make up a conscious sensation of what the body is suffering. When this mechanism is faulty a person who is perfectly capable of perceiving stimuli makes inappropriate judgements. The cases of patients whose amygdalae have been damaged are a case in point. Such patients seem not to be able to recognize the facial expression of fear in others and exhibit an uncommon friendliness to others they do not even know, which is not always wise because our very first encounter with strangers should normally cause a slight tinge of wariness, or even suppressed fear, until we can judge better what their intentions really are. This may seem an odd thing to say but watch very closely the next time you see two complete strangers being introduced to each other for the first time. Watch the distance between their bodies, the swift scrutiny of each other's faces and then the search for other clues in their comportment.

So, what we are talking about is the connection or association of emotional to cognitive behaviour. Perceptual data are melded into our faculties for emotional response and our constant grasping for meaning. And it is not necessary to suffer from a lesion in the brain to appreciate something of the loss of that connection. It can happen in a state of depression when the senses seem numbed, appetite suppressed, the world meaningless and life futile. This is because, owing to neurophysiological dysfunction, one is not joining what one sees or hears to emotion. This state would be alarming if one could detect alarm: music becomes noise; text on the page wearisome, devoid of charge or interest; people objects in movement demanding attention; social events tiresome and unpleasant; situation comedies on the television a refined torture; and work a tedious means of prolonging life. The depressive drags himself or herself automatically through the process of keeping fed and clean without any great conviction. This feeling may not be consonant with the objective facts of his or her existence – it can happen when he or she is employed and has a satisfactory family life. And yet this illness in the manifestation known as bipolar affective disorder, previously called manic depression, may be the downside or refractory phase of a heightened function of brain activity which, suitably channeled, leads to creativity and originality.

Emotions help bind humans together in society and contribute to the making of society in which, I argue, we are able to express our self-awareness. Man is essentially a social creature despite the counterintuitive evidence of war, violence and family disintegration. He (she) has an instinct for cohesion – the formation of groups for self-defence, task sharing and child rearing. Humans like forming friendship bonds, families, clans, football clubs, tribes and nations. Purely cognitive or machine-like communications are not comparable.

CHAPTER 5

In search of lost memories

In a recent paper in the *Proceedings of the National Academy of the United States of America* (*PNAS* for short) the word 'switch' is explicitly used.

Matthias Tradera and others say:

> recall requires a rapid and flexible *switch* from external perceptual reminders to internal mnemonic representations. (My italics)

Let's consider this by thinking about the English poet William Wordsworth.

In his infancy, meadow, grove, stream and lake were spread before him. He looked, heard, felt and drank deeply of experience. As he grew, the shapes and colours of solid external matter were transformed within him into another kind of richly-endowed inner substance – altered states of body and brain. In his later years, when far removed from lake and stream, he remained receptive to the modulations of this inner substance. The shadowy recollections of forms that had been imported long before now transcended current experience. With his 'inner eye' he could create a host of golden daffodils dancing in the sunlight by the edge of a lake.

When he was depressed he drew an inner strength from previous, transfigured experience. After the freshness and glory of childhood had passed away he drew comfort from the intimations of a distant immortal sea with mighty rolling waters which he had once been able to hear and which was his real home.

Between the intense receptivity of early childhood and later recourse to memories he could be unsettled by inner perceptions no less pale than the experience of an outer world.

> I was often unable to think of external things as having external existence, and I communed with all I saw as something not apart from, but inherent in, my own immaterial nature. Many times while going to school have I grasped at a wall or tree to recall myself from this abyss of idealism to the reality. At that time I was afraid of such processes. (Reference in the notes.)

Comparable states of mind are not granted to everybody and neither is the power to express them. Unknown to most of us, however, we too experience the muted transitions – in my terms switches – which constitute the play between the inwardly and outwardly directed operations of our thoughts and actions. The inward and the outward do not reflect simply the direction of processes but some sort of feeling about the characteristics of inner and outer matter, and also the nature of the signals between the states of matter. The inner is an intense conglomeration of biochemical activity which causes intense subjective experience. The outer is a substantial amount of matter composed differently. They are appreciated in two strikingly different ways.

Taking the first type of appreciation, which is common to most of us and on which I now concentrate, the process of inner creation, whether transcendently imaginative or banal, is forged by all of us and operates in the same way in everyman as in the visionary poet.

The inner substance, by the nature of its constitutive matter, dictated by forces of molecular shapes and transfers of energy, mediates processes different from the largely non-organic matter outside us. Although other humans are made of the same organic matter as ourselves (even though we only see surfaces such as skin and hair), and therefore carry out similar internal operations, we can only infer such operations by detecting what they export outside themselves through speech and other forms of communication.

Because we are essentially the same kinds of beings we exploit sensory signals to transform our own inner experience into states of internal matter which copy or mimic the internal matter of other humans. If we could not do so language* would not be possible.

> *Strictly, other types of animal communication are the norm but a separation between humans and other animals is justified to an extent because much of the former is genetically-driven association; in humans there is communication of imagination and creativity which transcends the genetically-programmed forms of communication.

A salient feature of internal processes, as revealed by inner experience, is a self-awareness which is fugitive in quality, as felt experience, but which plays within a cognitive certainty of the existence of a more durable inner substance that enables us to remember day by day who we are, our own name, and to speak, copy each other and play our part in society.

In order to experience the nature of outer matter, however solid, we can only seize it by inner processing within a different arrangement of matter. But the process of responding to immediate experience can be overtaken by more forceful inner recollection and so provoke doubts about the solidity of external matter. In those 'spots of time' where such perceptions passed through Wordsworth's awareness he did not or could not identify the provenance of the stimuli that were creating his awareness at those moments. Belief in an enduring aspect of his own identity was in danger. Internal experience had welled up in him with an uncommon force of inner apperception that coerced him into believing that he could rely solely on his own 'immaterial' nature. He could only find the means of rediscovering and re-connecting himself to the concrete external world by grasping a solid object.

Here I recollect a paper I cited in the Preface, by Luo and Maunsell, who refer to a 'lingering' effect, which I would call a 'subdued' loop ready to surface into consciousness, sought or unsought.

Wordsworth did not realize that the solid, coherent and orderly external phenomena were necessary in the first place to build what was internal through repeated exposure to reliable experience of the external. He did not realize, I infer, that the immediate experience of gazing at familiar scenery had the power to excite stored internal states corresponding exactly to what had been stocked by previous experience of the same shapes and colours. The act of looking, erroneously taken to be directed action uniquely outward, had started a chain of events, brought up from pre-existing internal states, overlapping in and out of synchrony with the operations provided by immediate experience. Both previous experience and immediate experience were bearing forcibly into him and he did not detect the difference.

These transformations of matter can overpower the more commonly enduring states of everyday experience, which most of us retain, that embody who and what we are. This poet differs from us because of the unusual intensity with which he is exposed to experience. But, unknown to us, we too go through similar processes when we revisit familiar objects because it is simply not possible *for us* (compared to the finest processes of neuroscience) to separate prior knowledge from immediate experience. The difference is that throughout the flux of sensations we do not usually lose ourselves.

Even if we are not able infallibly to distinguish between what we are remembering and what we are currently experiencing, most of us do not suffer from a misidentification of what we believe we are retrieving from memory and what we are currently experiencing.

There are two reasons for this.

The first is that in the normal course of living and behaving in the world such awkward questions rarely arise. The exigencies of the outside word shape the patterns of our survival, work and play. From an internal aspect, we do not seriously doubt that our physical actions have something to do with us and that a measure of persistence is constructed by some kind of inner psychological identity. At the same time, in everyday behaviour, we do not pay significant attention to the transformations of matter – even supposing we could reliably detect them – that mark the movements of our internal experience. But the fact is that when we are alert and looking at objects and scenes we are remembering more than we think we are, and when we are summoning cognition to supplement the act of looking we are detecting more than we consciously perceive or remember.

The second reason is that the brain has a mechanism in the hippocampus for separating current from recollected experience. But this mechanism can sometimes be faulty.

In short, Wordsworth was not correctly identifying the source of his experience. He had to take corrective action to recalibrate the position of his body by using tactile and gravitational data. Then his body consciously recognized that it was once more in the immediate presence of tangible matter.

Perhaps he was not mistaken.

To compose the story of the growth of his mind he thought deeply about its inner workings. He also knew that the external forms and characteristics of nature had played a crucial part in how he became what he was. His mind and his surroundings played against each other, and he was driven by the conviction that the entire universe was one substance. The soul of each human is the interface between the inner and outer dispositions of this substance:

> How exquisitely the individual Mind
> ... to the external World
> Is fitted: -- and how exquisitely, too
> ...
> The external World is fitted to the Mind.

An excerpt from a notebook encapsulates what the poet dedicated his life to understanding:

> To gaze
> On that green hill and on those scattered trees
> And feel a pleasant consciousness of life
> In the [impression] of that loveliness
> Untill the sweet sensation called the mind
> Into itself by image from without
> Unvisited: and all her reflex powers
> Wrapp'd in a still dream [of] forgetfulness.

I lived without the knowledge that I lived
Then by those beauteous forms brought back again
To lose myself again as if my life
Did ebb and flow with a strange mystery.

I have never said that our switches have the same solidity and certainty as a standard electrical switch for a lamp or piece of apparatus; I think of little episodes which I experience, so unclear and so fugitive that I cannot detect whether it is a memory or the memory of a dream, and which pass so fleetingly that I cannot grasp them.

Another type of appreciation

Although this chapter is mainly about one type of the sensory grasp of nature, there is another which is radically different and seems to be unique to humans.

This is the method of scientific enquiry and experimentation through complicated and refined measuring instruments which range from the humble microscope or thermometer to huge pieces of apparatus like the Large Hadron Collider and the James Webb Space Telescope. In between I put functional magnetic resonance instruments and other types used to investigate the workings of the brain.

Using these instruments, experiments should be replicable and theories developed to explain the data. Notably, human subjective bias should be eliminated to the extent possible.

I will return to this point in a short concluding section in the last chapter.

Chapter 6

A pine tree in the south of France and a lake in Switzerland

The skills of the great pictorial artist can induce or activate similar loops and trips in us, the viewers, so that, in a meaningful transaction, parts of the brains of the artist and viewer are brought momentarily into synchrony.

Some of the oldest art we know of are the famous cave paintings of Lascaux in France and Altamira in Spain, dated to roughly 35,000 years ago. (There are numerous other examples of rock art in most parts of the world.)

We do not know with certainty what their purpose was but as those in France and Spain mostly contain animals the artists were probably reflecting or celebrating their communities' reliance on them for food and clothing. They must have been made for viewing and perhaps helped bind groups together. Another possibility is the depiction of an animistic approach to what we may hazard to call the beginnings of a spiritual or 'religious' consciousness, in this case animistic. Even if that is so, we should not date such awakenings to the date above; such intimations could have started millions of years before, which is what I think.

Today it is noticeable how children, unprompted, love to draw and paint. Most will drop out when they realize they have no real talent. One man who did not drop out is my favourite painter, Cézanne, who said this about painting:

> … treat nature by means of the cylinder, the sphere, the cone, with everything put in perspective so that each side of an object or a plane is directed toward a central point. Lines parallel to the horizon convey breadth … Lines perpendicular to this horizon convey depth.
>
> Here's something that can't be disputed. I'm quite sure about this: an optical sensation is produced in our visual organ that makes us classify by light, half tones or quarter tones, the planes representations by color sensation. So light doesn't exist for the painter …
>
> Art is a harmony parallel to nature – what are we to think of imbeciles who say the artist is always inferior to nature?
>
> Painting after nature is not copying the objective, *it's realizing our sensations.* (My italics. Source referenced in the notes.)

Some have said that Cézanne and Mondrian were 'neuroscientists' but I think it better to reserve that term for scientists. However, it is clear that Cézanne thought deeply about how vision works more than 90 years before Hubel and Wiesel won the Nobel Prize for their ground-breaking work on how the eye and brain process lines.

The tree in a painting that I am thinking of is called *Large Pine and Red Earth*, painted on the Montbriant property of Cézanne's sister and brother-in-law.

Cézanne approached this painting with care and concentration making preliminary studies. He was very fond of trees. Joachim Gasquet, a poet and writer friendly with Cézanne, which was not always easy, (and sat for a fine, but unfinished portrait) wrote:

> He loved trees. Towards the end, with his need for sustained solitude, an olive tree became his friend ... The tree's wisdom entered his heart ... 'it's a living being', he said to me one day ... I love it like an old colleague ... I'd like to buried at its feet.

In fact, he was buried at the Saint-Pierre Cemetery in his home town of Aix-en-Provence. A mass was celebrated at Notre-Dame-de-Lorette in Paris.

I have a *Catalogue*, produced by Tate Publishing, UK, for a major exhibition of his works in Paris, London and Philadelphia. Of this picture it says:

> He viewed the pine frontally and from a closer vantage [compared to previous studies], making it the center-piece of a cross-shaped composition and emphasizing the patterns traced by the trunk and the branches emanating from it, whose tentacular forms have a sovereign vitality so powerful as to be almost disturbing ... the play of forms conveys the formidable energy of nature.

No photograph exists to my knowledge of this tree. We may be tempted to ask why paint a tree when a photograph would do just as well. But Cézanne did not set out to render the equivalent of a photograph.

I do not believe that a photograph can capture the energy of nature in the same way. When you look at Cézanne's numerous studies of trees with gnarled, forceful branches you can see that he must have both subtracted and added to his raw visual experience.

What does a painting do for us?

When a painting is made and then viewed I count *four* locations of states of matter that are related to each other. Switches are instigated in at least two of these states of matter.

The first state is the stuff 'out there' which initiated the act of painting. Note that I do not say 'trees', 'faces' etc. because these are *our* creations inside us and named in *English.*

The second is the stuff in the head of the painter who sets out to render what he or she perceives and wants to create on canvas, by which I mean all the loops and parallelisms which develop, play against each other and finally settle into more stable states that allow the artist to proceed and convey her or his experience.

The third is what is on the canvas. This bears a relationship to the first state of matter (the 'stuff out there') but will have undergone considerable modifications for two reasons. In the first place it is a result of the artist's inward thinking; in the second, there must be an essential modification by way of the fact that 'out there' for present purposes is a mix of organic and inorganic material of every possible combination whereas the canvas consists mostly of pigments and oils or similar substrates.

The fourth is in the head of a thinking and responding viewer who is prepared to dedicate time to his or her contemplation of the canvas and allow switches to happen.

> Cézanne has … attained so emotional and expressive a connection between the external world and his internal state of mind that one can truly speak of their temperaments as one.

My loop idea states that in a successful encounter with a painting, loops are set up in the viewer which correspond, to an attenuated and modified degree, with those originally in the brain of the painter. In addition, the viewer will bring his or her own experience of the subject matter into some degree of relatedness to the painter's intention.

It's important to look carefully at Cézanne's paintings and sketches, and make an effort to resonate with them. Words – as used here – are not up to the job; I can only nudge the viewer in a direction which I hope will be illuminating whilst at the same time conveying my obsession with this painter.

Let me quote from a critic:

> … how impoverished appear the words red, green and blue. It is easy to comprehend an impulse to prohibit them when one thinks about these mutes violets, these sumptuous greens, these strident reds, these dazzling oranges, this brutal, acidic blue … what's at issue are paintings true qualities, which … present themselves at every junction in an oeuvre that's nothing but painting and in which only souls amorous of painting can take delight …

Binocular vision

Normally we use two eyes to look but the images falling on both retinas are not the same because they are viewed from slightly different angles owing to how our eyes are spaced apart. (You can experience this simply by viewing an object fairly near by, opening and shutting each eye in turn. The object 'jumps' laterally from side to side.)

Therefore, an explanation is needed about how we see objects whole 'in the middle' so to speak. In effect the 'binocular rivalry' must be abolished or negated, leaving in the head a virtual construction of a complete and whole object, but calibrated with position of the external object as explained in Chapter 2. It is not entirely clear how the brain does this but we can surmise that the melded construct we experience with confidence, and are able to report, cannot exactly mirror or record the disparate data streams generated by each eye.

According to my idea, two distinctive loops are merged or reconciled with each other. The brain has made sense of excessive data and 'solved' the problem by creating an impression of a whole object. You could say that that is obvious but it is important to realize that we are not talking about solid stuff 'out there' but to a final act of creation entirely in the head, which is made of a different kind of matter.

The art critic Christopher Fry says this about Cézanne's struggles with form and colour:

> As the colour blossomed, successive contours narrowed down the zone of uncertainty round the form in rhythms that made visible the force of the organizing logic. The residual uncertainty that remained, a lateral vibration of edges, was perhaps the irreducible ambiguity of binocular vision.

Neuroscience teaches us that chromatic contrasts – i.e. different colours – can define the contours of an object although where colour is not conspicuous then lines may define the contours and this is accomplished by a *parallel* path.

Switch caused by choice of media used

Between 1878 and 1880 Cézanne made two renditions of the same scene in the Oise Valley; one oil on canvas and one with graphite, gouache and watercolour. This allows us to study what effects the two different media make.

In the first, as my Catalogue says, 'virtually all the brushstrokes align vertically'. Even the clouds are delimited by forms which are almost rectangular.

Now, as we make microsaccades from side to side *horizontally*, any *vertically*-aligned features will become more prominent. In contrast, in the predominantly watercolour rendition, although there are vertical strokes for a tree and a stripped bush, there is more a sense of the *volumes* of the spaces in the clouds in the sky and the fields in the foreground. The sensation is one of greater repose in the latter whereas in the former the sense is of greater immediacy, even a forceful insistence on the presence of the landscape.

Parallelisms and switches in the perception of colour

There is a beautiful painting by Cézanne of the Lac d'Annecy in Switzerland. It shows two buildings on the edge of a lake framed on the left by a huge tree. The canvas is divided horizontally by the line of the lake-shore in the middle. The artist achieves an extraordinary effect of the depth of the lake, which makes it seem as deep as the mountains in the background are high. But the overall effect is one of claustrophobia, yet pleasing although one critic has calls it 'menacing'.

What is most striking however is the colour scheme, principally in blues and greens with tiny flecks of yellow, orange and white.

Adrian Stokes, an English aesthetician, has written:

> It is a commonplace of colour 'science' that adjacent colours, such as greens and blues are difficult to 'manage'.

Cézanne was never one to respect such prescriptions.

Let me now concentrate more on colour. The M (magnocellular) and P (parvocellular) paths were introduced in Chapter 2, exhibited as parallelisms. The P channel is involved in our experience of 'colour', which indubitably happens in the brain but is causally created by the properties of external objects as they reflect certain wavelengths of light, depending principally what wavelengths are absorbed by the chemical constitutions of their surfaces.

Within 'colours' there are more parallelisms!

To explain: the *retinal ganglion cells* in the retinas of our eyes convert the photonic energy of light waves into chemicals which initiate the action potentials in neurons. The cones are for colour and rods for black-and-white vision.

The ganglion cells come in a variety of shapes, therefore function. There are ON-centre and OFF-centre cells, the first being excitatory and the OFF inhibitory. But why?

The answer lies in perceptual discrimination. Inhibited surrounding OFF-centre cells can enhance the gain of signal of the ON-centre cells. In other words, a comparatively weak signal from an object which has absorbed most wavelengths other than, say red, but may be less intense, can still be detected.

There are different sizes of receptive fields and response properties:

M cells can detect subtle contrasts over their large receptive fields and are likely to contribute to low-level resolution vision. P cells have small receptive fields that are well suited for the discrimination of fine detail. P cells and nonM-nonP cells are specialized for the separate processing of red-green and blue/yellow information.

Cézanne is playing, making us 'trip' against the common, inlaid loops in our head that engineer colour contrasts. At first glance it makes us uncomfortable but then, if we persevere, yields to a more satisfying experience because he has induced a trace in us – however faint – of the feeling which he experienced at the scene, or at least what he wanted to transmit.

In point of fact this place was a bustling resort and we may assume more colourful than depicted here.

Perhaps Cézanne was yearning for the light of Provence.

Chapter 7

Bagatelle

The power of music calls for comment. How can sounds produced by scraping bits of wire, blowing into tubes, or banging on leather and metal move one out of oneself or make the hair on the nape of the neck stand up?

Why do people enjoy social events dominated by music?

The subliminal emotion of waitingness, tinged with anxiety, lies at the root of our enjoyment of music when we voluntarily submit to a process of anticipation, climaxes or resolutions of themes. The brain is particularly good at adjusting when the unexpected happens, like a change of key, and in this brief transition of time, the body's emotional system is activated and made alert. Music therefore involves the physiology of alertness and resolution of associated emotions.

When some animals are frightened they erect the hair on their head to make themselves look bigger in order to confront an enemy or rival. The hair on the back of our necks stands up for the same reason.

Let us speculate: vocalized calls, supported by percussive effects or tones emitted from hollow sticks, became 'used' by our hominin ancestors to develop enriched communications skills which bound them together in more complex social patterns.

Humans are more effective in groups. By this I mean they were mastering their environment. In turn, richer social links and the inherited cultural artefacts further developed their intellectual powers.

The melody of a tune can also act as mnemonic device.

There are many reasons, I think, for our indulgence in music. We could look at a divergence between music and language. Language is a complex parcel of signalling methods often ranged round the concept of the self: the use of 'I' (as found frequently in this text), a continuous grounding in and reference to the body and conglomerations of beliefs and cognition centred on what one feels to be the enduring, organizing self-ness which acts and reflects. Music bypasses the constructed 'I', uncoupling the self from the stream of feeling, providing satisfaction in itself. When joined to ballet, we can admire shape, form and movement. When joined to dance, it is of course, intimately related to sex.

We can lose ourselves in music:

> For in music men let themselves go, in the belief that when they are concealed in music no one is capable of seeing them. (Nietzsche)

In songs, choral music and opera, music is more closely related to language but singing pathways in the brain are different to speaking paths though they overlap. Music also innervates the whole body producing a generalized but strong effect owing to a systemic hormonal change of state. Music permits realization of shades of emotion through changes in amplitude and of key much better than words, which are more restricted to a limited range of qualitative adverbs and adjectives.

Music is also essentially rhythmic, the rhythms established and re-enforced by loops. In choral participation or chanting at sports events the rhythm can help entrain equivalent states in others. In one way, each member of the crowd achieves little islands of togetherness. Drumbeats help to instil martial courage in a whole regiment. Rhythm sets up re-iterated and amplifying waves of neuronal states.

In the deepest recesses of the body, in parts seldom accessible to awareness, music brings to consciousness unsuspected worlds of meaning, unsought memories, and inexpressible shades and types of feelings. Music can retain the same theme and yet change it in limitless variations, as if echoing our progress through life, partly the same and partly changing until both come to an end.

Music can bring consolation and acceptance. What takes place through music is beyond the power of words to imitate or explain because what is brought to awareness is not images, nor thoughts, nor words, but traces of states and events dredged up from deep within the body in purest forms, coming into consciousness as perceptions and appreciations unmoulded and not capable of being moulded by words. It affords a sensation and experience dissociated from and uncontaminated by the self and by the appurtenances of words and meanings. It gives a shadowy glimpse, like the perception of movement at the boundaries of vision, into the mystery of being and existence.

In the context of the theme of this book, similar loops – of the brain and body enjoying music – are re-enforced in all of the hearer.

One kind of loop, and its distortions, that can be set up is when listening to a set of variations.

Take for example Beethoven's 'Diabelli' Variations whose basic theme is a trivial waltz but which Beethoven altered and embroidered with his mastery of forms of music.

Here I want to add that a variation makes no sense unless one loop is established retaining the theme against which play the subsequent variations. This argues for a continuity of the listener holding the theme – or loop – at least for the duration of the piece.

How does this work at the physiological level?

Sound is a force and works mechanically in the inner ear. A membrane in the cochlear called the basilar membrane is impelled into undulating motions by the pressure transmitted to a membrane of the cochlear whose force in turn has been transmitted from the tympanum, the eardrum. The motion of the basilar membrane sweeps hair cells called stereocilia to one side causing ionic channels in the hair cells to open, admitting calcium first which precipitates a chemical change in the cell with the release of a neurotransmitter inducing waves of excitation in nerves leading to the brain. (This is cranial nerve eight for the enthusiast.)

The notes in a minor key are placed less far apart in pitch than in the major key and probably require greater concentration to resolve; the basilar membrane may undulate less. To my disappointment I have seen no research on this. Vital is the fact that listening need not be passive; by concentrating we influence the stereocilia.

You reflect more when listening to a piece in a minor key, whereas when the tune blazes forth exultantly from a major key you could more readily surrender yourself to exhilaration, even dance. (The cochlear is right next to the vestibular apparatus in the ear which keeps our bodies balanced and with which it shares cerebral fluid.)

A composer may also use counterpoint – the simultaneous sounding of two or more melodic themes or harmonies that are related to each other but different. According to my scheme we could characterize this as the synchronous overlaying of loops.

But we should be aware that in other cultures other types of music predominate. Indian or Chinese music sounds strange to me but this must be a matter of enculturation. I do not believe that Western and Eastern ears are different. Accordingly, different loops of habituation must be established.

As an example of environmental or contextual influence we expect a funeral dirge to be funereal, sombre and serious. The effect will be gained by both tempo and key. An example is the funeral march from Handel's oratorio Saul.

***

Now to the title of this chapter by considering – better, listening to – that 'trifle', a bagatelle by Beethoven.

I am thinking of Opus 33, Six Bagatelles, No 3, Andante in E flat major.

This illustrates contrasts that music can create. When there is a contrast it makes you alert to the nature of the two types contrasted. In short, you have remembered one loop as you encounter a slightly different, but related, loop.

The piece starts with a rather bumpy, clattering theme alternating with more skittish passages until it relaxes into a peaceful 'singing', lilting mode which provides contentment and satisfaction.

The sixth piece, *Presto – Andante amiable e con moto*, descends into deep chords in the lower register of the piano. I find it difficult to explain why this moves me so much. It is as if a loop, waiting to be expressed has been dragged up out of the depths of one's subconscious. It evokes a similar feeling when listening to the third movement of the piano sonata, *Adagio sostenuto*, Opus 106 ('The

Hammerklavier'). About three-thirds way through this thoughtful, slow music the composer strikes a sequence of very low notes.

'Coming home'

Classical Western music is annotated by keys, e.g. B flat major. The system of keys is an artificial, imperfect system for approximating sounds to marks on a page through which we can compare and recognize discordances and concordances.

The composer may modulate between different keys. An example is the conclusion of Beethoven's Opus 109 which, having set up a restless state with alternating forte rushes, the piece finally arrives back, tranquilly, to its starting point.

Chapter 8

The cat sat on the mat

This profound sentence, so familiar to older adults from their days in primary school, heralds a topic about human behaviour of surpassing complexity, even surpassing the grasping of that cup of coffee.

I invite you to see past the meaning, which may strike you as insubstantial, and think about what is happening. Also, you will see that the switches in a good reader, at the mechanical level of reading, are imperceptible and will be more normal when we connect consciously with the meaning of the text.

The 'c', 's' and 'm' are *graphemes* which, in a process not fully understood, engender *phonemes* in the auditory cortex of the brain, that is discrete small units of 'sounds' which, after being joined up and passing through further hierarchical processing, confer meaning.

This might seem obvious but the whole point of this book is that such automatic, looping processes invoked to decipher a simple sentence like this conceal from us the astonishing fact that a 'c' on the page, as an abstract shape, can be converted into a sound, the only relationship between the two being a culturally engendered skill; that is, we must learn how to do it. You should then be free to meditate on how more profound and less familiar sentences from great literature operate.

If you are a good reader you are barely conscious of what is going on in your brain (and articulatory muscles) when you read; you 'see through' to the meaning. Only if I stop and ask you what grapheme(s) you have just detected will you separate your 'reading loops'.

What is more, as Stanislas Dehaene says, 'you have forgotten how you learnt to read' which in turn makes it hard for the 'normal' reader to understand why some people cannot do it. This is called a 'specific learning difficulty' but more often 'dyslexia'.

Further to explore the kind of complexity we are dealing with, and restricting ourselves to English, here is a contemplation of the 'th' sound.

There is the 'standard' sound, imperfectly transcribed as 'thugh'. But it may be said as 'thee' when you want to emphasize the following noun or simply give the sound itself more stress. Your brain deciphers these nuances.

In poetry and hymns, in order to get the right number of accented syllables in the poetic scheme the sound may, when needed, be shortened to a barely perceptible 'th' as in Milton's blank verse when required, which is commonly the metric scheme of the *iambic pentameter*, which has a pattern of ten syllables of which five are stressed.

Contrast:

> But O how fall'n! how chang'd
>
> From him, who in the happy Realms of Light …

In the second line count ten 'feet', with one for 'the' and two for 'happy' with

> What though the field be lost?
> All is not lost; th'unconquerable Will …
> And study of revenge, immortal hate,
> and courage never to submit or yield

'Th'unco-' makes two feet, not three.

noting also above that 'fall'n' rather than 'fallen' and 'chang'd' rather than 'changed'. (In a hymn or poem the last word sometimes has to be 'chang**ed**').

When a prosodic scheme like the iambic pentameter is deployed by reading over time as in an epic narrative, its incessant rhythm sets up what I call a 'foundational', base loop under the loops which sustain the overt meanings and associations, and this accords added power to our engagement with the poem.

For those who can read ancient Greek, it is said that the cumulative effect Homer induces, especially in *The Iliad*, is like that of a majestic cavalry charge.

When Johannes Gutenberg produced his press the typographers had to cast lead letters. In England Caxton did not have one to mimic the Old English 'Thorn' or þorn (Þ, þ), sound, so they joined 't' and 'h' which they had together, with hope, and this makes part of an *irregular* word, which must be learnt. (If the child has been educated in making the correct sounds for 't' and 'h' each on their own she will be thrown by the combination. Instead, the reader must memorize the sound(s) produced by 'th'.)

This causes no especial difficulty if the child has heard the sound enough times with subtle changes in emphasis because the different sounds are *generalized* to an 'abstract' principal sound, rather as we do not think that cat B is a different kind of animal to cat A because its colour is different or it is more fluffy.

The point of all of this is that the child hears these subtle variations but realizes that it is the same word. The loops are able to be inflected and modified just to the degree to allow them to handle variations but not so much that their core identity is lost. That said, if you are hearing for the first time somebody speaking your own language but in a savage dialect you might not be able to grasp the meaning the first time.

In short, specific loops are established to convert, say, 'c' to the sound 'k' (or in some cases the sound 's'). Then, I postulate, overlapping loops retain specific forms of meaning in the brain, first correlations between the graphemes and phonemes and then, after concatenations of phonemes, further transformations to arrive at the concept of 'cat'.

Note that 'c' may be soft as in 'face' when the brain, by training, sets up *this* loop. These loops are found principally in the frontal cortex, which is the associative or 'thinking' part of the brain but not *only* in the frontal cortex. The loops are not isolated in any part of the brain but must make further connections to the motor area of the brain which may have to produce the right sequence of muscular efforts to emit the right sound, and to the auditory cortex in order to have consolidated the sounds in the first place.

Accordingly, when reading, the 'bottom up' loops, which are activated by what the eyes detect, collide with or meld with the 'top down' loops established by training and memory. In this way the reader makes a safe connection.

Still staying with English, which is an alphabetical script in contrast to the logographic or ideographic system like Chinese which relies more on a pictorial representation of concepts, most experts agree that there are two paths for reading.

One is called the 'indirect' or 'phonological decoding' path and exploits the 'where' or dorsal path already explored. The second, the decoding path, by definition, is slower than the direct path.

Difficulties in reading

You may wonder how it is that 'normal' children can adapt to reading whereas it is extremely difficult for others.

First, scientists are not yet completely sure about how reading – and many other functions of the brain – work exactly. However, with modern techniques of brain imaging and other types of measurement they can trace the parts of the brain and interconnections involved in reading. Cognitive neuroscience is beginning to elucidate how the brain manages its operations.

We know that the brain is particularly good at finding patterns and significance out of seemingly chaotic and abundant streams of sensory data. What is also known is that the developing brain has an extraordinary degree of 'plasticity'; it can adapt, though there are limits. Also, plasticity or adaptability must be traded against the need to establish enduring loops which code letters to sounds reliably and fast.

In terms of the theme of this book we should think of some loops that become almost 'hard-wired', like those for walking; at the opposite extreme are those loops established when we listen carefully to a new piece of music, which may then become more solidified if we hear the piece again or have to play it ourselves. I hope you can see that the measurement of these loops, their intensities and persistence, will be very difficult. They will also vary from live performance to performance.

You may like to see a talk by Professor Stanislas Dehaene (see Notes) with some wonderful graphics showing what happens in the brain as the subject reads.

What we can say with confidence is that even in normal children a lot of explicit instruction and practice is necessary – very few learn to read by themselves, and it can take time because the child is doing something which is unnatural, that is converting abstract shapes into sounds which then have to be joined up and further converted into meaning. Talking is more natural because it evolved over millions of years.

But talking is very intriguing. The baby is primed to talk by inheritance – and equally the mother or carer – but the baby must be put in a talking situation with suitably adapted 'baby-talk', which is not babyish but a vital, wonderful human adaptation.

Reading phases

As said, learning to read is not simple. It requires practice under the supervision of a good teacher or parent.

Linnea Ehri, an American, is an acknowledged, widely cited expert.

She uses the word 'phases', not 'stages', to describe the development of reading skills. It implies that the student does not 'pass' critical stages but proceeds through phases with different emphases of processing and different levels of maturity. She asserts that the skilled reader uses at least five strategies:

- decoding – sounding out and blending graphemes into phonemes

- pronouncing common spelling patterns
- retrieving 'sight' words from memory (that is a vocabulary which must be built up through listening and reading)
- exploiting patterns, i.e. getting 'sight' from 'flight'
- predicting words from the context

In the fourth bullet above, the two words are distinguished only by the first grapheme. The notorious plaints about the irregularity of English orthography – foolishly argued by Bernard Shaw amongst others – ignores the patterns and regularities. If you commit '-ight' to memory words using this sequence should hold no fear. In my experience children who are good readers rather enjoy this challenge and will proudly display their knowledge of 'knight' and 'night', easily scooping up both the irregular grapheme ('-ight') and the silent 'k'.

Relevance of loop theory, switches and trips

If we single out the words 'through phases with different emphases of processing and different levels of maturity', it is easy to see how we may ground this in the physical reality of states of the brain where there are different loops, overlapping with each other, formed, activated and sustained at different times.

As we read, recognize and understand words, phrases, sentences, paragraphs and books, we pass through a succession of multitudinous switches and trips forming those intersecting, vastly complicated and intricate loops.

CHAPTER 9

To be or not to be

I separate live drama from literature because I think they operate in slightly different ways on us although in the end they work by inducing loops within us as close parallels to what the dramatist or author is conveying.

What is more, a recent paper by Emma Templeton and others reveal that fast response times to each other in conversation show a 'connectivity' between the parties, and so I extend this to live drama.

So, let us start with a scene which is the guard-platform of a castle. Two guards are on duty, one of whom complains that it is bitter cold and that he is sick at heart.

Two more characters enter and one says, 'What, has this thing appear'd again?'

The 'thing' is a ghost.

One of the men is sceptical but when the ghost does appear he is asked to address it because he is a 'scholar'.

We know we are not in Denmark. We probably do not believe in ghosts but this is how Shakespeare opens *Hamlet, Prince of Demark.*

In good theatrical performances, in spite of the artificiality wherein actors mimic being characters, you escape the loop which grounds you as a person and you surrender to believing that the Prince of Denmark is in front of you when part of you knows that it is an actor in a costume. You think no more about it and enter loops which evoke worries and undercurrents of feelings which could match your own: what is he to do when his uncle is occupying the bed of his mother? What is he to do about Ophelia whom he thinks he loves but treats badly and abandons to madness?

When a group of travelling actors arrive to perform a play – making a play in a play – he switches from his melancholy to a state of hypermania and excitement as he sets his trap, 'The play's the thing wherein I'll catch the conscience of the King'.

Once my tutor in literature asked me what the difference was between William Shakespeare and Christopher Marlowe. As I was unable to answer he answered for me. 'Consider the *groupings* of the characters in Shakespeare compared to Marlowe.' There are long monologues in Marlowe and no matter how beautiful his language one has more *interactions* between the characters in Shakespeare.

This was a clue for me: the interactions in a grouping of several characters or one-to-one induce loops, that is to say shared, behavioral or emotional states. We do not say these loops will be exactly the same; that is impossible because each of us will bring a different innate constitution and autobiographical history to the event but they will have aspects of similarity which, in effect, explain why drama appeals to most of us.

Let us leave aside uncles in the wrong beds and look at scenes from Act 1 in the light of my loop idea.

First, there are two guards; Francisco, and an Officer, Bernardo. Their mood is tired, cold, discontented and frightened.

Then enters another Officer; Marcellus, and Hamlet's friend Horatio, who is a sceptic (and later becomes a dramatic foil to Hamlet).

The first loops are those that belong to any dialogue: the speaker initiates a transaction that results in a shared experience in the hearer. They then convey to us in the audience the setting, which is cold and late in the night.

Unfortunately, this scene is often performed by second-rankers in a repertory company who simply set the scene but if it is well done we should be inducted into a feeling of foreboding, awe or wonder under an immense dark sky, waiting for the 'the bird of dawning', and made to start meditating on the nature of life, death and what, unknown, may lie after life.

In the next scene we switch to a warm interior with a grouping of a court.

Hamlet enters, morose and dejected.

Claudius, the King, makes a speech both statesmanlike and unctuous, passing smoothly over his hasty marriage to the late King's wife. He graciously gives Laertes leave to go to France but then refuses to let Hamlet depart.

The Queen says to Hamlet, 'Thou know'st 'tis common – all that lives must die, passing through nature to eternity.' There is a hidden loop, barely disguised, as she enjoys life and bed here on earth.

The King matches her platitudes, 'But you must know your father lost a father; that father lost his'.

In this complex grouping there are numerous interactions between the characters: the King preparing to defend his country and granting his favours; the obsequious courtiers, the queen regal but hiding her guilt; Hamlet, depressed without finding grounds for his depression except that he hates the King and thinks the wedding feast fell too soon after the funeral rites for his mother, though has no reason yet to impute any charge of murder against Claudius.

The court departs and Hamlet, left alone, delivers his first soliloquy.

The convention of the soliloquy is not common today but was frequently employed by Shakespeare. As far as we are concerned, the actor is exteriorizing his or her thoughts and can set up loops in us in the same way as a dialogue. We often talk to ourselves; we do not notice if a fictive loop takes control; we pay attention to what is being thought.

Hamlet delivers six soliloquies in all. The only other person, just once, is Claudius when he kneels to confess his sins, leaving his back exposed to Hamlet, who hesitates, offering pale excuses, and does not kill him. We cannot suppress a loop agitating our consciousness: 'Oh for goodness sake – just do it'.

Hamlet is depressed:

> How weary, stale, flat, and unprofitable,
> Seem to me all the uses of this world!

and cannot get it out of his head that the wedding was hasty and undignified.

He is completely stuck, but Horatio and two guards enter to tell him about the ghost. His excitement mounts.

Now we turn to another, contrasted grouping, a domestic scene with the King's Counsellor, Polonius, and his children Laertes and Ophelia. The father and son are much given to offering sententious advice to others.

These aphorisms can bore us and Shakespeare possibly goes on at excessive length. Notwithstanding that, the advice is often sound and offers rich pickings that find their way into anthologies.

There is an unpleasant tone: Polonius does not imagine that his daughter could be the bride of a Prince so he tries to steer her away from Hamlet.

We too can get stuck in silly loops turning round and round wherein we become enchained in our upbringing and the prejudices of our parents, schoolmasters and others around us.

Let's sample first Laertes to his sister, Ophelia:

> for on his choice depends
> The sanity and health of this whole state....,
> But, good my brother,
> Do not, as some ungracious pastors do,
> Show me the steep and thorny way to heaven,
> Whilst, like a puff'd and reckless libertine,
> Himself the primrose path of dalliance treads ...

Polonius is even more trenchant to his son and condescending to his daughter. We have all met this type, so certain in their views, not to be shifted from their own world view, with no sense of self-questioning and with never a doubt entering their heads. One wonders if he counsels his King wisely. We are quite glad when Hamlet kills him, noting here *en passant* that Hamlet manages to kill several people (at least five directly or indirectly) before succeeding in killing the one he is supposed to be killing.

> Neither a borrower or lender be
> This above all – to thy own self be true

To Ophelia, to beware of Hamlet, in spite of her protestations that she loves Hamlet and he her:

> Affection ! Pooh ! You speak like a green girl.

From a warm interior we switch suddenly back to that cold platform. We sense a climax but Shakespeare chooses to delay this with an episode which highlights the inadequacy of the King and Hamlet's disgust with him. Suddenly, cannons fire off. The King has ordained that a kettle-drum and trumpet proclaim every night his prowess in drinking.

The ghost appears and eventually tells Hamlet that he has been murdered by his brother.

Hamlet donates more sayings to a dictionary of quotations:

> thoughts beyond the reaches of our souls
> There are more things in heaven and earth, Horatio,
> Than are dreamt of in your philosophy.
> Something is rotten in the state of Denmark.

Between seeing the ghost and the return of the guards Hamlet delivers his third soliloquy, which includes:

> ... meet it is I set it down
> That one may smile, and smile, and be a villain.

He enters a state of hypomania and loses control of himself. The dramatist is going to show that this loop takes a downward turn as Hamlet becomes gripped again by doubts and languid inertia.

Too often I have seen Hamlet played steadily as a poised Renaissance prince. A close reading of the text makes it clear that at times

Hamlet is not mentally balanced; he suffers from a bipolar disorder. Shakespeare must have observed it in order to portray it with such accuracy.

We are dealing with innate, maladjusted biological rhythms that evade conscious control and adjustment. I maintain that 'loops' could be a partial explanation of this debilitating illness.

Back to the battlements, because there is a quite extraordinary scene. Hamlet seems almost mad. He is partly relieved to think he has the truth of his father's death (but sometimes suppressing doubts as to the ghost's veracity) yet quite wild in what he is going to do about it.

The first thing he tries is to ask Horatio and the guards to keep the revelations of the ghost secret.

Hamlet first says casually that they should just shake hands and depart.

Horatio:

> These are but wild and whirling words, my lord.

Then Hamlet asks them to swear on the hilt of his sword that they will keep quiet but is startled by the ghost's words from below, delivered four times, 'Swear'.

Now there is a moment of complete ambiguity.

> As I perchance hereafter shall think meet
> To put an antic disposition on.

There are several displays of this 'antic disposition' but is he acting or actually not quite right in the head? I incline to the latter view. A good actor could portray the ambiguity.

I will not say much about Hamlet's most famous soliloquy, 'To be or not to be', because it is straightforward in the expression of the fear of death, which although is seen as a release from life's burdens also worries us; we cannot be sure that we will really end and not be subject to alarming dreams.

> Thus conscience* does make cowards of us all.
>
> *I take 'conscience' to be equivalent to 'consciousness' and not 'conscientiousness'.

Now in a live dramatic rendition, this loop plays against another loop which we should be experiencing: why is Hamlet dithering so much over avenging his father's murder?

My loop theory can explain – to some extent. Many of us do hesitate, not certain of ourselves and what we should do. Others are more decisive.

Shakespeare sets up a clever contrast with the forthright, Norwegian Fortinbras who eventually bursts unto the scene. However, Hamlet is the more interesting character who forces us to think more about life.

And that is why we let our own internal loops play with those the author creates.

Chapter 10

More words, words, words

Polonius:

What do you read my lord?

Hamlet:

Words, words, words

Words are our principal method of communication, either for everyday transactions or through the deeper exploration of themes in worthwhile literature.

Yet there is a trap which often goes unperceived.

We are tempted to believe that words suffice, without our seeing their limitations.

For a start they are but tokens of communication – that is artificial constructs established by our culture – and are not always reliable. If you like, they are unreliable shadows of dreams of our true, inner feelings, and can be misinterpreted.

However, in what follows I am assuming that loops of reasonable similarity can be induced by good writing, and attentive reading structured on themes which resonate between writer and reader.

***

Earlier I treated live drama separately from 'literature' though some might construe the two as the same thing whereas I maintain that they work in different ways yet still create and use loops.

So now for the written word.

I am sure you have sometimes read a book or passage and fairly suddenly switched to the thought 'That's it!'. You are in tune with what the writer is expressing; you have entered a shared loop.

The choice of literature to choose from is vast so I must be selective.

Few of the authors I have read have explored consciousness or analyzed the transitions between consciousness and the less conscious as deeply as Marcel Proust.

Here is one of my favourite passages which shows mastery and psychological insight, together with a supreme ability in words. (Perforce a translation must be used here.)

> Then for a second time I create an empty space before it [an unknown state], I confront it again with the still recent taste of that first mouthful and I feel something quiver in me, shift, try to rise, something that seems to have been unanchored at a great depth; I do not know what it is, but it comes up slowly; I feel the resistance and I hear the murmur of the distances traversed.
>
> …

> Will it reach the surface of my limpid consciousness – this memory, this old moment which the attraction of an identical moment has come so far to summon, to move, to raise up from my very depths?

At this point I feel like giving up – it has been done better by Proust but let me make two points.

Notice how he re-tries to capture that memory lying anchored but capable of being unanchored from his sub-conscious. In my terms, he knows somehow that there is a loop there waiting to be accessed or loosed from its confines in the sub-conscious and ready to effect a switch to awareness.

Second, you may admire the metaphor but not be brought into a loop, because you might protest that too much is being made of dipping a little cake into an infusion of tea, which is the famous event Proust describes and which ignited his memories and urged him to write about them after living a dilettantish life.

I myself enter this kind of loop when memories involuntarily surface of fishing in deeper water on a lake in Africa. The hook, bait and weight were far below, invisible; a sudden, thrilling tug on the line; we hauled in, imaginatively giving an identity to the catch. An indistinct, flashing shape was observable until nearer the surface the shape and the colours it contained took on the form of an identifiable fish.

Proust's memory is from his childhood in a village called Combray. And he finds a wonderful analogy to explain how indistinct formless shapes come into distinct existence.

> And as in that game in which the Japanese amuse themselves by filling a porcelain bowl with water and steeping in it little pieces of paper until then indistinct,

> which, in the moment they are immersed in it, stretch and shape themselves, colour and differentiate, become flowers, houses, human figures, firm and recognizable, so now all the flowers in our garden and in M. Swann's park, and the water-lilies on the Vivonne, and the good people of the village and their little dwellings and the church and all of Combray and its surroundings, all of this which is assuming form and substance, emerged, town and gardens alike, from my cup of tea.

and later:

> The great modification brought about by awakening is not so much our entry into the clear life of consciousness as the loss of all memory of the slightly subdued light in which our mind had been resting, as in the opaline depths of the sea.

Proust reflected on the outer world and an inner consciousness and realized that through his art as a novelist he could share experiences:

> ... my thoughts ... in the depths of which I felt I could bury myself and remain invisible even when I was looking at what went on outside ... When I saw any external object, my consciousness that I was seeing it would remain between me and it, enclosing it in a slender, incorporeal outline which prevented me from ever coming directly in contact with the material for ... [from reading] came the emotions aroused in me by the action in which I would be taking part, for these afternoons were crammed with more dramatic and sensational events than occur, often, in a whole lifetime ... It is true that the people concerned

> in them were not what Françoise [the maid] would have called 'real people'. But none of the feelings which the joys or misfortunes of a 'real' person awaken in us can be awakened except through a mental picture of those joys or misfortune ...'

In my terms, he is switching between loops, not necessarily deliberately, and I postulate that there must be different loops for outwardly and inwardly directed attention.

I turn now briefly to Tolstoy, another favourite author.

His most prominent loop was his fear of death, counterbalanced by other loops in which inhered an acceptance of the vividness of life which can be enjoyed.

Another series of loops lie in his descriptions of marriages which he dissects with an unerring, pitiless eye.

Let us contrast two households. One is the family life of Stepan, Anna's brother in *Anna Karenina*. He is fun-loving and lovable but is unfaithful to his wife, most recently with a ballet dancer, and everybody knows it. The household is thrown into disorder as Dolly becomes more and more unhappy.

Tolstoy neither begins nor ends his novel with Anna. The beginning starts with Stepan and Dolly, the end with another character, Levin, and his anguished search for the meaning of life, even though *he* is happily married.

I think we all know such loops though they will be coloured and inflected differently because our experiences are different. However, there is no point in a writer attempting such descriptions and characterizations if she or he cannot induce near-equivalent effects in the reader.

Let me develop this issue of outwardness and inwardness which I adumbrated in the chapter on memories.

The outer world before me now has inert objects made of wood, metals, plastics and textiles. Outside the window are hills, trees, fields and other animals as well as man-made objects and structures. The world outside me and the world inside me are composed of sub-atomic particles and atoms that obey the same physical laws. There are no atoms inside me that cannot be found outside me.

But there is a big difference between the two worlds. Inside there is a concentrated mass of sophisticated molecules unique to life, in which carbon is a common element, making carbohydrates, proteins, fats and nucleic acids. The molecules bend and twist presenting different profiles to other molecules, bonding with them and separating from them, exchanging protons, storing and releasing energy, modifying the distribution of electric charge as ions fluctuate rapidly across membranes, metabolizing, circulating between compartments in cells which grow, divide, move, specialize and die. These molecules create an inner substance with its own special qualities. As J.D. Victor says, there is a 'mapping from events, perceptions and actions to something very different: patterns of neural activity'. Complex interrelated events in the brain develop where conformations of matter take shape as a result of input from the outside as well as its integration with internal transfers of memories and cognition.

Now imagine two individuals, each within his or her private internal world but also communicating with each other. This communication between humans induces similar processes in each party as some conformations of matter follow and internally mimic the output of each speaker in alternation. When people speak to each other they do not simply hear, process and respond; they adjust and modify their internal tissues to match, anticipate and align their own experience to what is being heard and processed. They transfer behavioural patterns to each other and mutually influence physical changes in each other by bringing them into a degree of correspondence.

The tokens of communication become concretized in social and other types of network which are essential to maintain their existence. Examples are societies, religious groups, legal institutions, places of instruction and learning, and computer networks.

The exterior world is now populated with new things which in turn will provide new stimulation, bearing in on the inner world and requiring yet greater powers of internal cognition and learning in order to understand the very same outer world that its ancestors and others have created.

The pursuit of science allows better and more objective understanding of this exterior world. However, the individual cannot yet measure inner subjective experience with the same degree of certainty. For the time being the novelist and poet may better capture the essence of these loops.

Proust reflected on the outer world and inner consciousness and realized that through his art as a novelist he could explain the link between the two.

Chapter 11

Eureka!

In Greek, this word means, 'I have found it!'

It was supposed to have been spoken by Archimedes when he hit upon a method of determining the purity of gold. Sometimes the call is transferred to the moment when he leapt out of his bath having struck on the principle that bears his name which is that the weight of the water displaced by his body equaled his own weight, which provided the principle of buoyancy or upthrust, explaining why iron ships can float.

Examples of trips in most of this book have been derived from the arts or humanities but in science there is also creativity and imagination, which I take to be yoking two or more loops together in an unlikely or serendipitous manner. I can only explain this by supposing that either closely located neurons, the 'grey' matter and/or long association fibres, the 'white' matter of the brain also known as fascicles, facilitate communication between areas of the brain. When this happens, I infer, loops with no apparent relationship to each other come into contact and establish a state in the brain that is powerful enough to reach consciousness.

Concerning a possible difference between creativity in the humanities and in science, Malcom Bowie said this about Proust:

> [He] writes with admiration about the slow, patient procedures of the scientist and places that slowness in perpetual dialogue with the happy accidents and intuitive leaps which also fuel intellectual and artistic work.

I am not sure I agree. Charles Darwin does seem to comply with the conventional assessment but I do not think it is true for all science. Equally, in thinking about pictorial artists and writers, we should not ignore all the years of apprenticeship when loops were being formed.

However, it is not possible to build a secure case for all forms of trips or, to be precise, about when these trips intercepted the loops; I can only be suggestive.

One thing to consider though is whether trips may be completely fortuitous or the result of the patient building of loops. I think I can show both types.

Both of the following instances may be apocryphal but this does not negate the fact that originality was involved. It is of no great moment because we may assume, without fear of contradiction, that even if the scientists had partly made up the stories, and indeed *because* they might have made them up, they were searching consciously for a way to explain the workings of their sub-consciousness, or 'out-of-awareness', even if they did not use or recognize those terms.

It is also worth remarking that any loop that 'comes to life' is likely to have been installed there in the first place by long and intense thinking around the problem.

Hence we are advised 'to sleep on a problem' and let night-time operations like dreaming take over. In such states loops suppressed by day may come to the fore precisely because they are not restrained by the daytime loops of purposeful and habitual activity.

A falling apple

When Newton had to go home to Lincolnshire to avoid the plague in Cambridge he was sitting in his garden when he saw – or was hit by – a falling apple.

He then connected the fall of the apple with the motion of a planet, each constrained by the force of gravity but obeisant to the law of momentum.

A snake swallowing its tail

Kekulé was a German organic chemist who said that he had thought of the structure of benzene, after a dream or reverie.

Benzene has six carbon atoms linked by alternating single and double bonds, each carbon also bonded to a hydrogen atom.

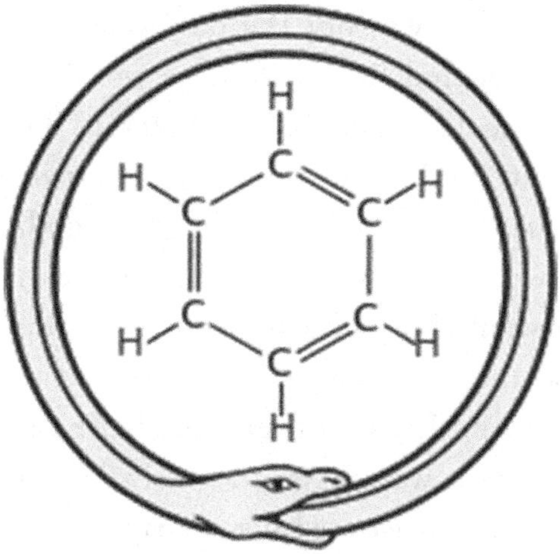

The unexpectedness of the link he made consists in the fact that, whilst thinking about carbon structures, he claimed he had been dreaming of the ouroboros, the snake swallowing its own tail.

On the origin of species

It seems that Darwin did not have such a single eureka moment, rather a long, slow dawning of an idea as he observed the differentiation of species on islands and, from his studies in geology, gained an intuitive understanding of the enormous stretches of time implicated. We cannot be sure. I do not think that his notebooks reveal a single moment of revelation.

He also knew that pigeon fanciers and dog breeders could produce new characteristics and behaviours after a few generations. In fact, he kept pigeons.

I have to remind myself constantly to think about time when I go to the Natural History Museum, London, and look at the variety of species from whales to insects, finding it scarcely credible that they all evolved from a common ancestor.

Furthermore, Darwin was not the only one who thought about natural selection. Alfred Russel Wallace working in the Far East had had the same idea and sent a letter to Darwin who was disconcerted but, with true honesty, acknowledged Wallace's insight. He determined to pull his thoughts together in a book but first made his own summary. It was arranged that his summary and Wallace's letter would be read together at a meeting of the Linnean Society. And so these two Englishmen avoided the mean dispute between Newton and Leibnitz over precedence in the invention of the calculus. Darwin was slightly reclusive and not a combative man and this slightly frustrated his colleagues who lent him their full support.

Our great scientist preferred his domestic life in Kent and his 'thinking path', a chalky route around his house.

On the Origin of Species is a 'slow burner' of a read, with just one illustration. It piles detail on detail until it forces its overwhelming argument.

It was an instant success on first publication and its first printing ran out on the first day.

It was not an accident. The academic world in this field, including Huxley, Lyell and Hooker, and scientists as far away as Germany knew that something was brewing in Darwin's head and were also responding to incessant demands by correspondence for more information on subjects where he knew he knew less.

In short, I characterize this not as a sudden trip but the outcome of innumerable loops which consolidated intense and deep knowledge playing against each other in an awesome dance of cerebral play.

Einstein

In the pantheon of great scientists we have to list Albert Einstein.

I have not studied him as much as Darwin and therefore cannot say what his thought processes were leading up to the extraordinary year of 1905 when, aged 26, he published four important papers.

Chapter 12

The disenchanted loom

I have deliberately broken apart the loop which underpins 'The enchanted loom', a phrase coined by the notable physiologist Charles Sherrington, writing about the brain and spinal chord as he knew it in the early 1900s.

As for us now, we have almost reached the end of our exploration of loops and parallelisms, and I find them less enchanting – though fascinating – than a mixture of the predictable and unpredictable, the promising and the worrying. Where I do not find enchantment also is in the thought of the entrenched loops in humans which militate against sensible, far-sighted and less selfish actions.

As a short reminder of some themes, and by way of example, there are parallel paths for picking up a cup of coffee, and for reading. There are loops implicated in everything we think or do. There are manifold, overlapping loops which work at different speeds.

My idea echoes this sentence by Semir Zeki:

> The intersection of two [parallel paths] and/or momentary dominance of one loop over another gives us the fleeting impression of consciousness or self-awareness.

Substitute 'switch' or 'trip' for 'intersection' and you have the basic idea because the important point is that we have little or no conscious control over these trips, which is why I have also chosen these terms, implying events that take place accidentally or by pure chance.

Zeki's quotation at the start of his book has this wonderful excerpt from T.S. Eliot:

> ... There is, it seems to us
> At best, only a limited value
> In the knowledge derived from experience.
> The knowledge imposes a pattern, and falsifies,
> For the pattern is new in every moment
> And every moment is a new and shocking
> Valuation of all we have been.
>
> *Four Quartets, East Coker*

This is only partly true as I hope I have shown.

Can you encourage – consciously – entry into a particular loop?

Yes!

For example you may wish to know how novelists have treated marriage. You pick up *Anna Karenina* if you are in a gloomy mood or *War and Peace* if in more optimistic humour.

This might seem as if I am contradicting myself by implying conscious control over a choice.

Not so!

Consider what hidden loop activated this very desire to think more about marriage – its triumphs and tragedies – and to know more about how novelists have treated the subject.

In short, we are not conscious of which loops create further loops but we may be able to console ourselves with the thought that we can perhaps control or overwrite our subconsciousness to some extent.

Avoidance of chaos

It may be objected that, if we are constantly swayed by inadvertent trips or a mish-mash of loops, how we ever get anything done. How can we possibly function at all?

> We are free to abandon ourselves to one or other of two forces, one arising in ourselves, emanating from our deepest impressions, the other affecting us from without. (Proust)

Proust captures a little of what I am trying to say but for the preliminary pronoun, 'we'. Better would be to have said 'the body is abandoned to …'. This will partly be addressed below but here we can say that our very enculturation makes us move most of the time in a purposeful direction.

If, as a child, you have been brought up adequately you will observe your father and mother going to work regularly in order to keep your house warm and put bread on the table. You yourself will be encouraged to get up, clean your teeth, dress yourself and go to school. Of course, now and then you squeal and object, and if you are afflicted by a precocious existential crisis may wonder 'why do all this?'

But there is more to add: these strivings cannot wholly be attributed to culture. In man there is an instinctive, combative drive to explore, endure, take risks, try new ways of doing things and, if necessary, traverse huge distances to discover richer resources or escape from a poverty of such. Those innate drives which also created culture have brought us to where we are today, for all its benefits and evils.

> ... the later a particular brain region matures, the less it is shaped by genes and more by the environment. (Sapolsky)

The 'self'

This notion has caused all sorts of bother in philosophical discussions from those like Hume who, notoriously, could not find it, to religions like Buddhism which deny its reality.

Hume's argument holds no force for me. I think his fundamental error is to try to find it by a method of introspection using language. That is wrong because you have willingly entered a loop – a loop pre-eminently in language mode – which is precisely where you should not be.

The Buddha's argument is more subtle and is centered on an ontological speculation, which is that there is nothing more than a sequence of 'flame moments' rather like a flame which passes along a stick.

> When a flame burns a piece of wood, one says that it wanders along it; nevertheless there is nothing but a series of flame-moments. Likewise there is a continuous series of processes which incessantly renews itself, and which is falsely called a human being.

This has merit but still commits several errors. Following Markus Gabriel, it is false to say there is no human being, even if we give it the wrong name or its existence seems to be an illusion. Illusions actually *exist* even if they muddle us.

To be sure, the 'self' is not some sort of master controller in the middle of the brain or an identifier similar to a rating plate on a washing machine.

Before even thinking about a 'self' I root my argument in the fact that we have a body and this body does all manner of complicated things holding loops, switches and trips together.

Another point: '*We* have a body' is a questionable sentence in that it implies there is a self which possesses a body. Better framed: 'there *are* bodies'. We can see them all around us and we ourselves can see a lot of own bodies. I attempted to draw out the consequences in the chapter about picking up a cup of coffee, and where I argued, with many others, that the prolonged making of and use of tools expanded our perception of ourselves.

That is just for vision and action but, as several commentators have observed, this body is also conspicuously subject to all kinds of experiences, principally internal, which I have tried to sketch.

It could help if we employ the passive or impersonal voice in our discussion. We could also eliminate 'I' or 'we'. There is nothing logically wrong with this sentence when you are Joe Brown and say, 'Joe Brown would like to come to the cinema with you', but it would be construed as an affectation. 'I' is more economical as long as we do not attribute to it a substantive amount of tissue with closed borders. The holding of and generation of a vocal action follows from the fact that 'I' sits in a vastly complicated network of brain tissue. Again, following Gabriel, it is incoherent to say that there is nothing corresponding to 'I'. It may take us a long time to discover the approximate locations and diffused boundaries of these complicated strata of loops but that is no reason why it cannot be done.

Therefore, not '*I* decided to reach for a cup of coffee', rather 'an arm was seen (or felt) to reach for a cup of coffee'. In this case, the body before the coffee was not surprised to see an arm suddenly extended in front of it. And, in most cases, the body can attribute a reason for this experience; it felt like another sip of fortifying liquid.

Suppose it is not a cup of coffee in question but a bar of chocolate. The body has taken a small mouthful with a degree of feel-good intention and in its consciousness decided to defer the rest of the consumption to later, mindful of unnecessary greed or its waist-line. Suddenly the body 'comes to' or there is a trip when the body realizes that the arm has shot forward and automatically picked up the rest of the bar of chocolate.

So not '*I* felt such-and-such a sensation' but 'a such-and-such sensation *was* felt somewhere, somehow, and there was a body which must have underpinned it and was able to report it'.

To be sure there are loops in which 'I', 'me', 'my', 'we' and 'us' are embedded that allow the articulation of 'I'm here' and so forth to be economically generated but this does not require us safely to define what the 'I' is, and where it is precisely.

Culture and civilization

Let us consider again what loop could have led one to think about the subject of 'marriage'.

There may be two loops that led you to think about this matter. First, your own is in trouble and you want to learn more. Second, your tutor in literature has asked you to write an essay about how Tolstoy treats marriage.

But the tutor herself is also the 'victim' or vessel of cultural knowledge and the progress of civilization, including her own university, which has been created by man and of which there is nothing quite like it in the rest of the animal kingdom.

At the same time, animals do have their own cultures and research is beginning to show how complex they can be and also how often hidden from us.

There is one important fact about humans, which is the extraordinary prolongation of our childhood in which we learn, in a kind of virtuous cycle, how to handle the very thing we have created, for example, reading, which builds on drives which other animals do have – that is an urge to communicate. Reading then builds a literature from which we can further learn.

'A brain in a vat'

This phrase relates to worries expressed by some, and touched on in the Preface.

How do we know that there is a 'real' world 'out there' and we are not brains in vats arranged by a malevolent alien power which has stuck electrodes into us in order to produce our experiences including those of an external world we appear to scan?

We can't know for certain but two points are in order.

First, that is exactly what we are!

The brain in its skull is bathed in cerebro-spinal fluid which surrounds it, retained by three membranes (the meninges), and in the spinal chord. It also fills a substantial system of four ventricles.

Access to the outer word, both to receive sensory data and to effect actions – mostly through muscles*, for example, to talk – are through nerves that pass through this vat.

*Modern techniques developed to help the paralyzed produce muscular action through capturing electrical 'brain waves' or wishes or intentions to move.

The brain also releases hormones that can effect physiological changes.

Second, the postulation of a malevolent power requires the existence of a 'real' outer world because a real entity, with powerful 'real' tools for inserting micro-electrodes, *must* exist outside our brains. If not I cannot see any logic in the arrangement.

I urge we abandon the words 'real' and 'reality'.

Constructive and non-productive loops

Most of this book has given numerous examples of constructive or productive loops.

I should repeat that I am not talking exclusively about the brain. One example worth citing is the complex looping in the spinal chord which enables us to walk rhythmically, one foot alternately in front of the other with hands in synchrony (left foot forward, right hand back etc.). This necessitates a complex pattern of activation and inhibition. The brain plays little part in this but can intervene to avoid an obstacle or deliberately impose a different pattern of locomotion to achieve a particular, more conscious objective.

Alas, there may be non-productive or destructive loops.

In the dreadful disease of Alzheimer's the patient can get stuck in a loop and will go round endlessly on the same track much to the exasperation of the carers. Doctors and psychiatrists call these 'fixations'.

Another cognitive loop I can think of is where a person remains stuck with his or her political point of view no matter what evidence that is brought before him or her that an opposite opinion is more rational or more concordant with the facts. At the time of writing I can cite denials about man-induced climate change even as fires, floods and other catastrophes ravage the planet. Yet another example is vaccine hesitancy or refusal even as millions are dying.

Of baboons, trolleys and cream cakes

Baboons

Enter baboons, preceded by a hirsute researcher, with an unusual profile and who can out-hair any known animal, and who has studied them extensively.

This man is Robert Sapolsky, a professor at Stanford University in the USA, who holds two posts which reflect a dilemma he was confronted with when trying to choose his academic specialty.

In 2003 he was a professor of biological sciences at Stanford University and of neurology at Stanford's School of Medicine.

He is also a research associate at the National Museum of Kenya.

In the first capacity he researched the behaviour of primates in the wild in Africa; in the second the science of the brain where more quantitative approaches are preferred, in other words the use of accurate, reproducible measurements. Speaking in an interview on the web site *Edge*, he said he was not sure if he had succeeded in reconciling the two approaches.

Baboons display mammalian behaviour in regard to care of the young but also considerable aggression as the 'alpha' males struggle for domination.

But Sapolsky found and repeatedly stresses that alpha male behaviour is not always as effective as exploiting strategies of social cohesion and being less domineering over females.

> A typical male baboon is too impulsive and can't possibly do the disciplined thing. Baboons are far less disciplined than chimps and when you map their brain anatomy you notice that they don't have a lot of frontal cortical function … they typically blow it at just the right time … The whole world is three seconds long – they're very pointillist on their emotions … they can play chess in their social landscape almost as well as chimps in terms of moving the right piece around, but at the critical moment they simply can't stop themselves from doing the impulsive thing.

This is from observation. Measurement in the laboratory has so far yielded fewer conclusive proofs. We do know a considerable amount in detail about how neurons and hormones work but that corpus of knowledge falls far short of explaining the complexities of social behaviour.

Sapolsky himself does not put it this way but I think that we have two separate factors to contend with. The first is the complexity of the human brain and its development, the second the complexity and variety of cultures and environment. Their interaction is not fully understood. Richard Lewontin, a geneticist, has said that we may never be able to disentangle the two.

A trolley on a railway line

I now illustrate a parallelism.

You are standing beside the points on a railway line where it divides into two tracks. Directly ahead you see a group of five people working on the line; beyond the bifurcation you see one person on a line. (This isn't the parallelism.) Suddenly, behind you, a heavy trolley is hurtling to the points. Do you let it run straight ahead and likely kill five or do you throw the points so that trolley veers sideways and kills one person?

A second scenario is proposed. Would you deliberately push a person onto the tracks to stop the trolley?

This ethical dilemma has been debated endlessly as a 'thought experiment' by philosophers and others. Neuroscientists can image the brain experimentally as subjects consider what they would do.

Sapolsky, attributing Josh Greene, says:

> Contemplating pulling the lever, and dlPFC [dorsolateral Pre-Frontal Cortex] predominates, the detached, cerebral profile of moral reasoning. Contemplate consigning the person to death by pushing them, and it's vmPFC [ventromedial Pre-Frontal Cortex] (and amygdala), the viscera profile of moral intuition.
>
> … a relatively minor variable determines whether people emphasize moral reasoning or intuition, and they engage *different brain circuits* in the process, producing radically different decisions. (My italics)
>
> In short, we may be able to act by 'thinking', or we may, benevolently, give way to a natural, inbuilt, aversion to deliberately taking a life.

Away from trolleys and tracks, what strikes me is that context, personal make-up and even whether or not you had a good night's sleep could play quite arbitrary roles in what circuit your brain 'chooses'. This makes me fear for our ability to tackle pressing problems like climate change.

To close out this little section, let me relay a cogent point Sapolsky makes which hits home neatly if I am honest with myself.

> **Going easy on ourselves also reflects a key cognitive fact: we judge ourselves by our internal motives and everyone else by their external actions.** (My emphasis)

Thus, I deplore people flying excessively round the world enjoying multiple holidays and contributing too much carbon to the atmosphere. But I have the bad feeling that I myself could be doing more with the application of more discipline in avoiding that car trip or better regulating the central heating. Can I unloop and deny myself that avocado from Peru or rasher of bacon from Denmark?

Cream cakes

You put a cream cake in front of a child of three years and say that 'if you wait 20 minutes without eating it I will give you another one'.

The child usually takes the cake but most children of about five or more can manage to defer immediate gratification for a greater reward later. A more powerful loop has been instantiated in the cortex which is able to judge greater reward for more self-control.

A provisional rounding up

I submit that adventitious trips between loops, or switches between parallelisms, create a momentary illusion of changes of perception experienced in the alert body.

I attribute these trips to happenstance, which in the moment you imagine there must be some mysterious 'consciousness', an entity that has so far eluded a precise definition acceptable to all and therefore has to subject to indirect and tangential research.

However, all that is happening in the mature brain is not totally subject to chance. I have argued that our enculturation, education and training including aspects of moral behaviour usually encouraged initially by good parenting – all partly 'inventions' of civilization but also, confusingly, instinctively driven – impose patterns of behaviour which for the most part have been instrumental in the preservation of the species.

Whether it is enough to navigate ourselves through the multiple crises currently confronting humanity remains to be seen.

Speculation

Some time ago I started another file called 'Bioluminescence'.

Two of the many papers and one good short book filed there are:

The Synthesis and Function of Luciferyl-adenylate and Oxyluciferyl-adenylate W. C. RHODES AND WILLIAM D. MCELROY From the McCollum-Pratt Institute and Department of Biology, The Johns Hopkins University, Baltimore, Maryland (Received for publication, May 28, 1958)

Hirano, T., Hasumi, Y., Ohtsuka K., Maki, S., Niwa, H., Yamaji M. and Hashizume, D. (2009) 'Spectroscopic Studies of the Light-Color Modulation Mechanism of Firefly (Beetle) Bioluminescence' (JACS Articles, *J. AM. CHEM. SOC.* 131)

Wilson T. and Hastings, J.W. (2013) *Bioluminescence – Living Lights, Light for Living* (Harvard University Press)

One diagram intrigued me. It is of 'chromatophenes' (coloured phosphenes) in plate 20a of Samuel Zeki's book, *A Vision of the Brain*, 1993, Blackwell Scientific Publications. Frustratingly, Professor Zeki could not tell me where K. and G. Beckers were. Presumably these pictures were produced by the subjects by hand in response to subjective experience responding in turn to stimulation such as transcranial magnetic stimulation.

Confessedly, what follows is complete speculation.

Suppose that in the brain there are molecules like luciferin and an associated enzyme. According to the energies generated by specialized molecules in the brain different 'colours' could be generated.

For a place in the brain, one could look first at V4 but it could also be in the inferior temporal cortex (the 'lower' path explained in Chapter 2), or even in the association (frontal) cortex.

Then the sensation of 'colour' could be created in the brain, which adds to the overall state of the brain. I am not proposing a *picture* in the brain. That would lead to an infinite regression.

However, this state of the brain could be aligned with spatial data which is integrated by the superior colliculus (and/or posterior parietal cortex) to give the misleading impression that the quale offered to perception resides 'out there' whereas it is 'made' in the brain, because it cannot be made anywhere else.

Different considerations would apply to audition, olfaction, pain, etc.

Notes and references

Preface

Science writer Guy Cox has also noticed this.

Ch 1 - *De profundis*

Occipital lobes and LGN

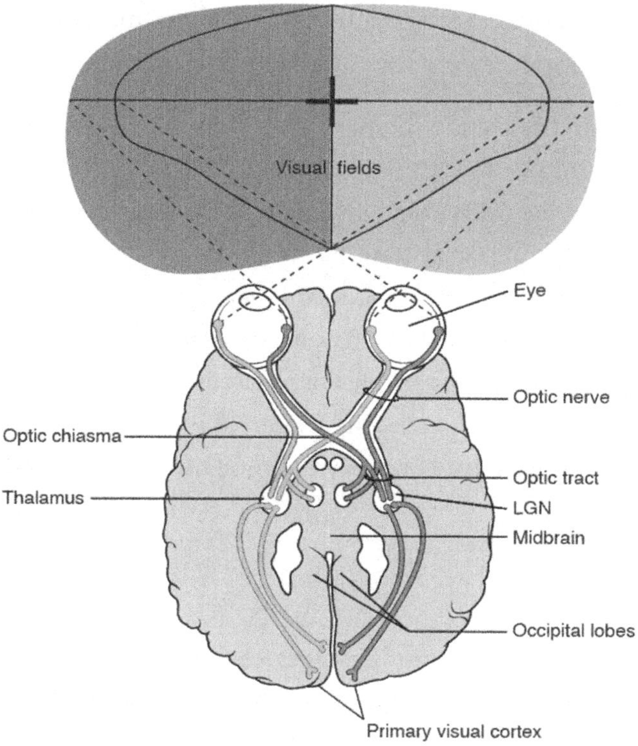

In the visual system signals are projected from the eyes to the *back* of the brain from where signals are projected *forwards*, increasingly modified and structured in hierarchies, to the front of the brain, the 'association' or frontal cortex. More about this in the second chapter.

For the sheer complexity of the brain you might like to sample this paper:

> 'Cerebral cortical folding, parcellation, and connectivity in humans, nonhuman primates, and mice', David C. Van Essen and others
>
> *PNAS* | December 26, 2019 | **Vol. 116** | no. 52 | 26173–2618

and also this paper in *Science* (2022), 375 by Luke Campagnola called 'Local connectivity and synaptic dynamics in mouse and human neocortex'.

To quote from the section 'Dynamic flexibility in the cortical circuit':

> Most studies in cortical synaptic physiology describe the circuit in a quiescent state … In effect, the cortical circuit is modified from moment to moment based on the recent history of activity, so it may be misleading to conceptualize the cortex as a single static circuit … these patterns [explained in the paper] suggest an ability to dynamically *switch* between different network modalities … (My italics)

Proust

I came across these intense passages in Proust when he had been contemplating the death, in pain, of his grandmother. This novelist, who was the son and brother of a medical doctor, evinced considerable knowledge about medicine and psychology, and more than a passing knowledge of the science of the day.

> World of sleep – in which our inner consciousness, subordinated to the disturbance of our organs, accelerates the rhythm of the heart or the respiration, because the same dose of terror sorrow or remorse acts with a strength magnified a hundredfold if thus injected into our veins: as soon as, to traverse the arteries of the subterranean city, we have embarked upon the dark current of our own blood as upon an inward Lethe meandering sixfold, tall solemn forms appear to us, approach and glide away, leaving us in tears'. (**Vol 4**, *Sodom and Gomorrah*, translated by Malcom Bowie in *Proust Among the Stars* (1998), Fontana)

'Subterranean city' is not a metaphor I would have chosen but works as an image of a world invisible to us.

> And then, on sleep's chariot, we descend into depths where the memory can no longer keep pace with it and where the mind stops short and is forced to turn back. Sleep's horses, like those of the sun, move at so uniform a pace, in an atmosphere where no resistance can any longer arrest them, that some small alien aerolith is needed … if this regular sleep is to be affected … and made to return, wheeling suddenly about, towards reality … From these deep sleeps we then awake in a dawn, not knowing who we are, being nobody, quite new, prepared for anything, our brain finding itself emptied of the past that had hitherto ben our life. … Then, from the black storm through which we seem to have passed (*but we do not even say we*) we emerge lying prostrate, without any thoughts: a 'we' it

> might be without content. What hammer-blow has this person or thing that is here received that it should be aware of nothing, be stupefied, up until such time as the memory comes hurrying back, and restores its consciousness or personality? (My italics)
>
> ... We possess all our memories if not the faculty of recalling them... But what is a memory we cannot recall?
>
> ... for I realized [on waking up] that hitherto all I had done was to dream that I had rung [for the valet]. I was alarmed to think, however, that this dream had the clarity of a cognition. Could cognition, by the same token, have the unreality of a dream?
>
> ...
>
> If it is an idea that sleep has forged, this quickly disintegrates into tenuous, irretrievable fragments.
>
> *In Search of Lost Time*, **Vol 4**, *Sodom and Gomorrah*, tr, John Sturrock, 2002 (Penguin Books)

Stringer *et al.*

Stringer, C., Pachitarin, M., Steinmetz, N., Reddy, C.B., Carandini, M. and Harris, K.D. (2019) 'Spontaneous behaviors drive multi-dimensional, brainwide activity', *Science*, **364**

Toker *et al.*

Daniel Toker and others (2022), 'Consciousness is supported by near-critical slow cortical dynamics', *Proceedings of the National Academy of Sciences of the United States of America*, **Vol 119**, No 7 (Open access)

Oscillations

Alejandra Sela and others (2021) 'Increasing and decreasing interregional brain coupling increases and decreases oscillatory activity in the human brain', *Proceedings of the National Academy of Sciences of the United States of America*, **Vol 118**

Underwood

(2021) 'A Sense of Self', *Science*, **Vol 372**, Issue 6547

Gonzalez *et al.*

Gonzalez, W.G., Zhang, H., Harutyunyan, A. and Lois, L. (2019) 'Persistence of neuronal representations through time and damage in the hippocampus', *Science*, **365**, 821-825

Ch 2 - To see or not to see

These are the principal books I consulted:

Zeki, S., (1993) *A Vision of the Brain* (Oxford Blackwell Scientific Publications)

Kandel, E.R., Schwartz, J.H. and Jessell, T.M. (1991) *Principles of Neural Science* (Appleton & Lange)

(There is a more recent edition.)

Hubel, D.H. (1995) *Eye, Brain, and Vision* (Scientific American Library)

Here is a simplified diagram of the surface features of the brain.

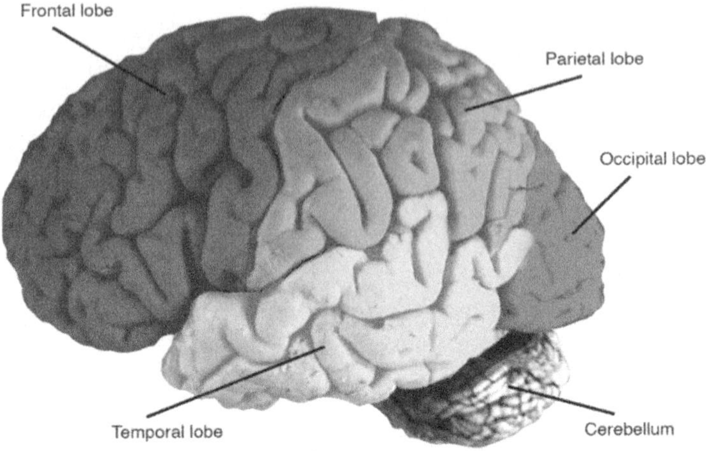

Note the huge size of the frontal lobe, which distinguishes humans.

But the cerebellum has more neurons than all the lobes above it.

Zeki

All areas of the cerebral cortex send outputs to more than one area. From which it follows parallel outputs from all cortical areas. These parallel outputs are not just to other cortical areas, but to subcortical centres as well. For example, all visual areas send fibres to the superior colliculus, a midbrain structure, and to the pulvinar, a subdivision of the thalamus ['fore-chamber', principal way-station for sensory input into the cortex (except olfaction).]

> ... Parallelism also implies a distribution and since all cortical areas have parallel, it follows that each distributes its signals to several other areas and, hence, it is common to speak of the cerebral cortex as consisting of several *parallel-distributed* systems ... [My italics]
>
> Chapter 13, *A Vision of the Brain*

May

May, P.J. (2014) 'Superior Colliculus', *Encyclopedia of Neurological Sciences*, **Vol 4** (Elsevier)

This structure is not diagrammed in this book but you can find it in Wikipedia.

Livingstone and Hubel

Livingstone, M. and Hubel, D. (1988) 'Segregation of Form, Color, Movement, and Depth: Anatomy, Physiology, and Perception', *Science*, **Vol 240**

Fissures

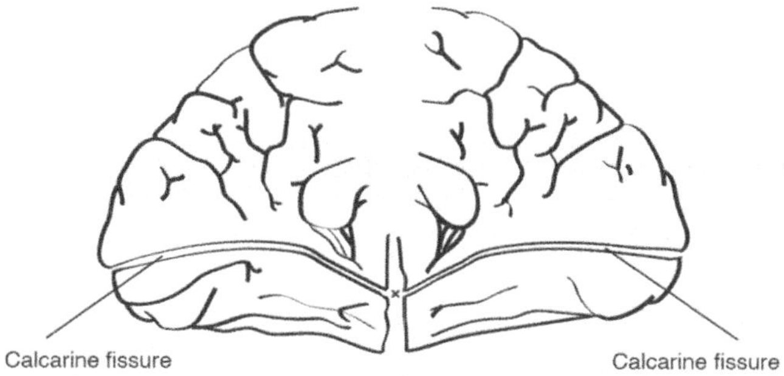

View from the back of the brain. The point marked by the little cross corresponds to the mid-point of our visual field. There are no cells at that point, only cerebro-spinal fluid. So the brain has work to do to make a 'whole' house in front of you. (However, the eyes also move across the scene and make sure that visual data falls on a receptive part of the retina which then reaches real tissue in the occipital tissue.)

Vision paths in more detail

Movement, moving shape and depth:

Retina (*rods* and large ganglion cells)

LGN (*magnocell* layers)

Occipital cortex

V1 (layers 4Cα*, 4B, 6)

V2 (*thick stripe*)

V3

V5

Lobes

parietal

(some *temporal*)

Form, depth and some colour:

Retina (*cones* and small ganglion cells)

LGN (*parvocell* layers)

Occipital cortex

V1 (layers 4Cβ, and 2 and 3 *(blob* and *inter-blob*))

V2 (pale *interstripe*)

V4

Lobes

temporal and some parietal

Colour:

Retina (*cones* and small ganglion cells)

LGN (*parvocell layers*)

V1 (layers 4Cβ, and 2 and 3* (*blob* and *inter-blob*))

V2 (*thin stripe*)

V4

Lobe

temporal

*Most of the cerebral cortex consists of six layers of distinct types of cell. But when V1 was mapped in more detail layer 4 was seen to be further divided. Instead of renumbering from scratch extra layers were added to layer 4.

Davidoff and Roberson

Davidoff, J. and Roberson, D. (2002) 'Development of Animal Recognition: A Difference between Parts and Wholes', *Journal of Experimental Child Psychology* **81**, 217–234

Crick

Crick, F. (1995) *The Astonishing Hypothesis: The Scientific Search for the Soul*

(Amazon)

I confess I have not read the whole book but a quotation from it in another publication.

Goldberg and Wurtz

I regret I cannot remember where I saw this. It may be:

Goldberg, M.E. and Wurtz, R.H. (1972) 'Activity of Superior Colliculus in Behaving Monkey - Effects of Attention on Neuronal Responses', *Physiological Reports*, **Vol 35**, Issue 4, 560–740

Ch 3 – On picking up a cup of coffee

'Reference frames'

After I had written this chapter my son drew my attention to Jeff Hawkin's book *A Thousand Brains*, which has many points of similarity. Also, I did know about the work of O'Keefe and the Mosers in respect of place cells and grid cells.

Manipulations of our hands

The human wrist can exert a power grip, doing what no other animal can, by laying the implement across the palm and holding it with every digit. The four fingers are placed round the object, say the haft of a stick or hammer, with the thumb braced over the haft opposite the index for a haft of smaller diameter or with thumb and fingers partially encircling a wider haft. In a precision grip for threading a needle both objects are usually held between the tips of the thumbs and the underside of the tips of the first digits. To hold a paintbrush, the second digit is sometimes added to thumb and first digit to support the brush. A less precise and less forceful grip is made by holding the object, a key say, between the pad of the thumb and side of the index finger. A hook grip for picking up a briefcase by the handle involves four bent digits. There is both a disc grip using all five digits to unscrew the lid of a jar and a spherical grip to hold a ball or stone. A lateral power grip, to pick up a heavy book, uses the thumb on one surface and two, three or four digits on the other surface. There can be combinations of grips like unscrewing the cap of a fountain pen or small container with one hand, and complex manipulations using both hands such as tying shoelaces. The fingers can punch for typing and they can point.

Goodale

Goodale, M.A. (2014) 'How (and) why the visual control of action differs from visual perception', (*Proceedings of the Royal Society B*)

Deloison

Deloison, Y. (2004) *Préhistoire du piéton – Essai sur les nouvelles origines de l'homme* (Plon)

Donald

Donald, M. (2001) *A Mind So Rare* (Norton)

Neuronal control of hand movements

Takei, T., Confais, J., Tomatsu, S., Oya, T. and Seki K. (2017) 'Neural basis of hand muscle synergies in the primate spinal cord', *Proceedings of the National Academy of the United States of America*, 8 August

Wilson

Wilson F.R. (1998) *The Hand – How its use shapes the brain, language, and human culture* (Pantheon)

Tool use and language share syntactic processes and neural patterns in the basal ganglia

SIMON THIBAULT
HTTPS://ORCID.ORG/0000-0002-7491-9488
RAPHAËL PY
HTTPS://ORCID.ORG/0000-0002-0678-0795
CLAUDIO BROZZOLI
HTTPS://ORCID.ORG/0000-0002-1719-8506
+7 authors Authors Info & Affiliations

SCIENCE•12 Nov 2021•Vol 374, Issue 6569•DOI: 10.1126/science.abe0874

Tools

Cambridge Encyclopedia of Human Evolution (1992), edited by Steve Jones, Robert Martin and David Pilbeam (Cambridge University Press)

Flintknapping

Whitaker, J.C. (2004) *American Flintknappers – Stone Age Art in the Age of Computers* (University of Texas Press)

Ch 4 - Green-ey'd monster

Two loops exemplified

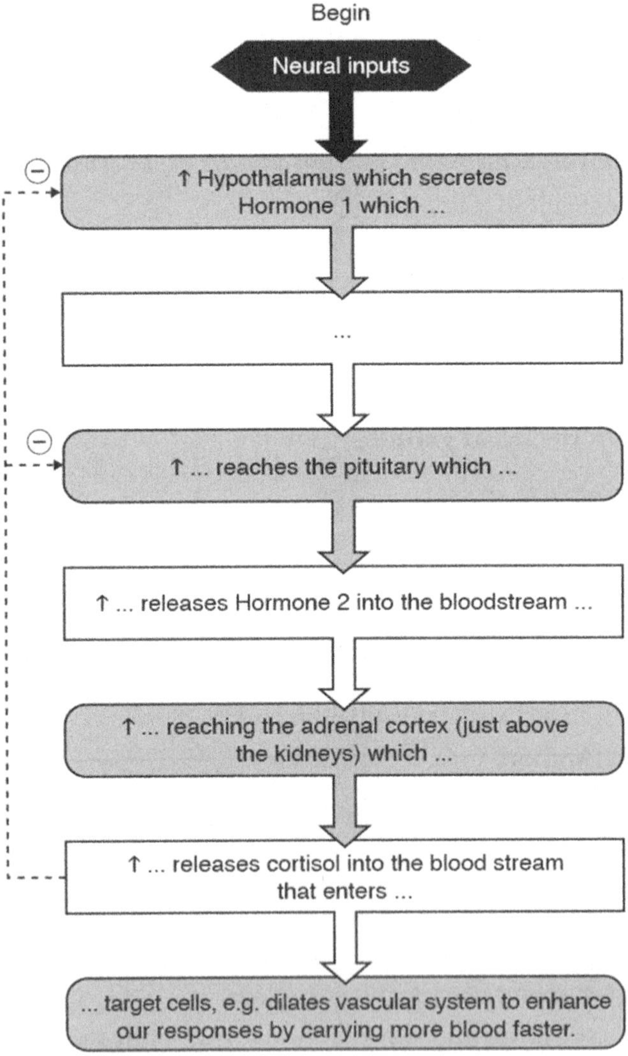

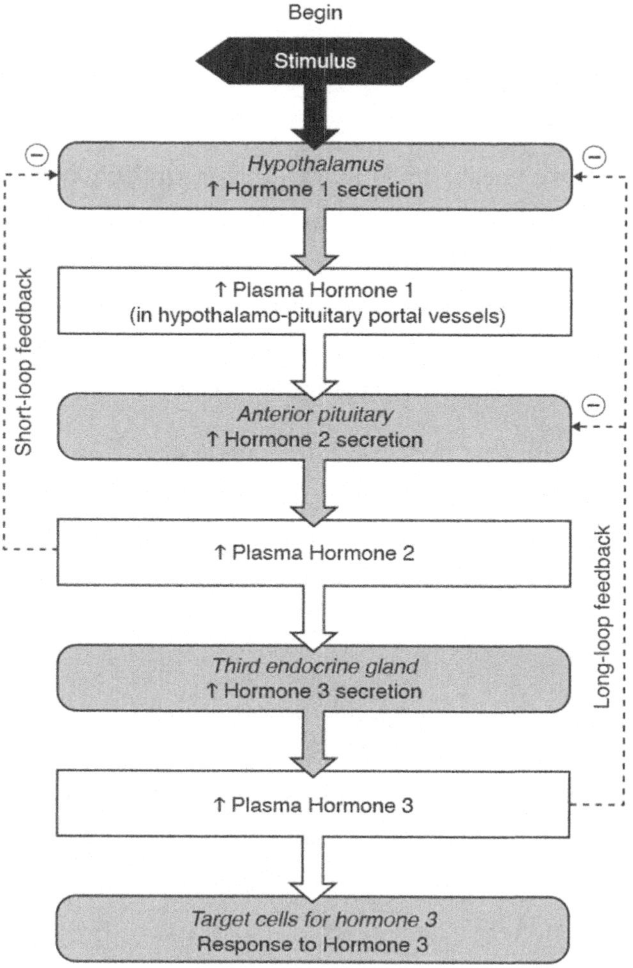

Adapted and slightly simplified from Vander's *Human Physiology* ('portal' means connected by blood vessels)

LeDoux

LeDoux, J.E. and Brown, R. (2017) 'A higher-order theory of emotional consciousness', *Proceedings of the National Academy of Sciences of the United States of America* (published online February 15, 2017) and

LeDoux, J. (1998) *The Emotional Brain: The Mysterious Underpinnings of Emotional Life*

See also 'Feel the beat', 15 January 2022, Caroline Williams, *New Scientist*

> In recent years, it has become clear that to really understand mental health, you need to factor in just how much the brain cares about what is going on below the neck …
>
> The reason that we generally aren't cognizant of this life-sustaining system [the parasympathetic] is because interoceptive messages are *below the radar of conscious awareness most of the time.* (My italics)
>
> …
>
> Some scientists believe that the brain's integration and interpretation of them is what provides the sense of self, that there is one 'I' living in this moment and experiencing the world in real time, and that any change from the status quo has an important impact on whether that experience feels good or bad.
>
> …
>
> The challenge to unravelling this is that the brain-body pathways involved are fiendishly complicated. The body is crammed full of sensors, not to mention kilometres of nervous pathways connection the organs to each other and the brain. Body and brain communicate in both directions …

Hagihara *et al.*

Hagihara, K.M., Bukalo, O., Zeller, M., Aksoy-Aksel, A., Karalis, N., Limoges, A., Rigg, T., Campbell, T., Mendez, A., Weinhotz, C., Mahn, M., Zweifel, L.S., Palmiter, R.D., Ehrlich, I., Lüthi, A. and Holmes, A. (2021) 'Intercalated amygdala clusters orchestrate a switch in fear state', *Nature*, **Vol 594**, 17 June

Ch 5 – In search of lost memories

William Wordsworth

The Prelude or Growth of a Poet's Mind (Text of 1805), with notes, edited by Ernest de Selincourt (Oxford University Press, 1936, revised by Helen Darbishire 1960)

This was not the title given by Wordsworth himself. It was his wife's after his demise. Wordsworth's own, more prosaic title was, *Poem – Title not yet fixed Upon - by William Wordsworth Getty addressed to S.T. Coleridge.*

Ch 6 – A pine tree in the south of France and a lake in Switzerland

Cave paintings

With regard to pigments, you might like to read *The Brilliant History of Color in Art* by Victoria Finlay, The J. Paul Getty Museum, Los Angeles.

She says that some of the pigments might have been transported a distance.

Catalogue

Cachin, F., Cahn, I., Feilchenfeldt, W. Loyrette, H. and Rishel, J.R. (1996) *An Exhibition Catalogue*, Trustees of the Tate Gallery, London,

UK, with Galeries Nationales du Grand Palais, Paris, France, and the Philadelphia Museum of Art, USA

First reference: Letter to Emile Bernard, April 15, 1904

Second reference: To Emile Bernard, December 23, 1904

Third reference: To Joachim Gasquet, September 26, 1897

A portrait of Gasquet is in the Národní Galerie, Prague, Czech Republic

Large Pine and Red Earth

The Hermitage Museum, St Petersburg, Russian Federation

Internal and external states

> Catalogue, op. cit., p. 468, on *Mont Sainte-Victoire Seen from Les Lauves*

Impoverishment of words

> Catalogue, (op. cit.) p 32, quoting Thadée Natanson in *La Revue Blanche*, 1895

Proust

In Search of Lost Time, **Vol 1**, *The Way by Swann's*, translated by Lydia Davis (Penguin Books, 2002, first published in French in 1913 as *À la recherche du temps perdu, Du côté de chez Swann*)

Binocular vision

Catalogue (op. cit.) p. 75, Christopher Fry, Preface to the watercolor and pencil drawings exhibition organized by the Northern Arts Council of Great Britain, 1973

The Oise Valley

The Oise Valley, Oil on canvas, Private collection, c. 1880

In the Oise Valley, graphite, gouache and watercolor, The Metropolitan Museum of Art, New York, USA, 1878-80

Lac d'Annecy

1886, The Courtauld Institute Galleries, London, UK

M and P cells

The quotation is from *Neuroscience – Exploring the Brain*, Bear, Connors and Paradiso (2001) 2nd Edition, (Lippincott Williams & Wilkins)

Another book well worth reading is *Eye, Brain and Vision*, David H. Hubel (1995) (Scientific American Library)

Ch 7 – Bagatelle

Definition

This is from *The Concise Oxford English Dictionary* (2011)

Key

Explained in Wikipedia (accessed 09-08-2021)

In music theory, the **key** of a piece is the group of pitches, or scale, that forms the basis of a music composition in classical, Western art, and Western pop music.

The group features a tonic note and its corresponding chords, also called a tonic or tonic chord, which provides a subjective sense of arrival and rest, and also has a unique relationship to the other pitches of the same group, their corresponding chords, and pitches and chords outside the group. Notes and chords other than the tonic in a piece create varying degrees of tension, resolved when the tonic note or chord returns.

The key may be in the major or minor mode, though musicians assume major when this is not specified, e.g., 'This piece is in C' implies that the key of the song is C major. Popular songs are usually in a key, and so is classical music during the common practice period, around 1650–1900. Longer pieces in the classical repertoire may have sections in contrasting keys.

This web site directs you in turn to others, for instance:

From http://www.wu-wien.ac.at/earlym-l/logfiles/em.2001-03

'Diabelli' variations

33 Variations in C [major] on a Waltz by Anton Diabelli, Opus 129, so a comparatively late work

(*Für Elise*, is the *Bagatelle in A minor*, Opus 59)

Saul

HWV 53

Ch 8 – The cat sat on the mat

Iambic pentameter

A type of metric line used in traditional English poetry and verse drama. The term describes the rhythm, or meter, established by the words in that line; rhythm is measured in small groups of syllables called 'feet'. 'Iambic' refers to the type of foot used, here the iamb, which in English indicates an unstressed syllable followed by a stressed syllable (as in *a-bove*). 'Pentameter' indicates a line of five 'feet'. (*Wikipedia*, accessed 07-09-2021)

There is a short discussion of this issue, comparing Italian and English, by C.H. Sisson who translated Dante's *The Divine Comedy* for the Oxford World's Classics.

The use of the English iambic pentameter for poetry incurs risks. If not done skilfully you will produce 'tumpty-tum-ness'. In my view, John Milton avoids this by modulating carefully – and powerfully – between patterns of stress and flow. I think that T.S. Eliot metricates very accurately.

On the other hand, although he achieves some striking effects by challenging the norms, I do not think that the poetry of Gerard Manley Hopkins is in English.

What is the relevance of metrication to hearing (and thus reading)? It shows that the nurtured brain, within limits, can master variation.

Dehaene's YouTube talk

How the Brain Learns to Read

https://www.youtube.com/watch?v=25GI3-kiLdo

Phases in learning to read

Ehri, L.C. (2002) 'Phases of acquisition in learning to read words and implications for teaching', in *Learning and Teaching Reading*, ed. Stainthorp and Tomlinson (The British Psychological Society)

Ch 9 – To be or not to be

Templeton *et al.*

Escalator effect

Templeton, E.M., Chang, L.J., Reynolds, E.J., LeBeaumont, M.D.C. and Wheatley, T. (2022) 'Fast response times signal social connection in conversation' (*Proceeding of the National Academy of Sciences of the United States of America*, January 25, 2022, 119 (4) e2116915119; https://doi.org/10.1073/pnas.2116915119)

Social intelligence

In a recent review in the journal *Science* called, 'Towards collective animal neuroscience – Social intelligence requires the study of both groups of brains and the individual brain', Julia Sliwa summarized two papers, also recent, by Rose *et al.* and Báez-Mendoza *et al.*:

The study of social interactions in neuroscience has been carried out by recording neural activity in the brains of two individuals simultaneously and led to the discovery that brains in a group of two *can couple their activity.* (My italics)

This review was published in the edition of *Science* of 22 October 2021, **Vol 374**, Issue 6566.

The two papers cited in the same journal are:

Rose, M.C., Boaz, S, Yartsev, M.M. and 3 authors (2021) 'Cortical representation of group social communication in bats'

and

Báez-Mendoza, R., Mastrobattista, E.M., Williams Z.M, and 2 authors (2021) 'Social agent identity cells in the prefrontal cortex of interacting groups of primates'

Shakespeare

I am using the Tudor Edition of the Complete Works, edited by Peter Alexander. January 1962 (Collins)

Dictionary of quotations

There is a famous story of a person watching this play being performed who objected that it was only made up of quotations!

There are over 300 quotations from *Hamlet* in *The Oxford Dictionary of Quotations*, 1959.

Ch 12 - The disenchanted loom

Sapolsky

2017, *Behave – The Biology of Humans at Our Best and Worst* (Bodley Head), and interview on the website *Edge*

About the author

Stewart Sanson graduated in English Language and Literature from the University of Oxford, UK. Currently he is a student with the Open University, UK, completing his degree in Natural Sciences. Before that he studied neuroscience at King's College, London, for two years.

He worked as a volunteer for the charity Dyslexia and Literacy International for 16 years.

This book is a shorter version of *More than a dream*, available on Amazon Kindle, which in turn is a shorter adaptation of a book called *More than a dream - Human experience and the behaviour of organic matter* which has not been published and is more than 200,000 words long. It has more about my childhood and its disturbing mental events which led me to take up these studies.

Made in the USA
Monee, IL
04 November 2026

67379927R00085